COMPUTATIONAL
EXPLORATIONS
IN
MAGNETRON SPUTTERING

COMPUTATIONAL EXPLORATIONS IN MAGNETRON SPUTTERING

E. J. McInerney

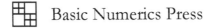 Basic Numerics Press

ISBN: 0692289925
ISBN-13: 978-0692289921

Basic Numerics Press

San Jose, CA
www.basicnumerics.com

To my daughter Rebecca

TABLE OF CONTENTS

Preface...ix

1. Introduction...1
   1.1 What Is PVD Sputtering?..1
   1.2 Why PVD Sputtering?..2
   1.3 Magnetron Sputtering..3

2. Magnetic field...9
   2.1 Magnetic Charge approach...9
   2.2 The B field of a Single Magnet—Numerically..............13
   2.3 Analytic solution...15
   2.4 Arrays of magnets...18
   2.5 Pole Plates..28

3. Electric field...33
   3.1 Ion Matrix Sheath...34
   3.2 Child's Law Sheath...37
   3.3 Two Fluid Model..43
   3.4 Magnetized Sheath...50

4. Electron Motion...53
   4.1 Motion With Two Magnets...53
   4.2 Drift Velocity...64
   4.3 Varying Magnet Strength...70
   4.4 Full Magnetron...75

5. Collisions..81
   5.1 Collision Types..81
   5.2 Probability of Collision..82
   5.3 Cross Sections...84
   5.4 Energy Change...90
   5.5 Velocity Change...91
   5.6 Anomalous Bohm Diffusion...96

6. Erosion 99

   6.1 Trajectories with Collisions 99

   6.2 Single Electron 103

   6.3 Erosion Profile 108

   6.4 Full Target Erosion 113

7. Deposition 117

   7.1 Radiation Analogy 118

   7.2 Discretization 119

   7.3 Deposition 123

   7.4 Uniformity 125

Epilogue 129

Appendix A. Mathematica Decoder 131

Appendix B. Cross Sections 133

Appendix C. Speeding up Mathematica 135

References 139

   Bibliography 139

PREFACE

"When the experiment is continued for some time, a dark deposit is formed on the glass around the extremity of the platinum wire, giving an extended conducting surface..."

W. R. Grove, 1852

In 1852, while experimenting with vacuum discharges, William Grove deposited silver on a glass substrate by Physical Vapor Deposition (PVD) (Grove 1852). In the century and a half since this discovery, PVD sputtering has matured into a common large-scale industrial process used for making thousands of products.

Grove's path into PVD was a purely experimental one. We are going to take a different route. Through the use modern computers and math software, we will explore the physical principles behind PVD sputtering. By creating small models and simulations of different portions of the sputtering process and experimenting with these models, we will build our understanding and intuition for the underlying physics of these systems.

There are many fine math software programs that could be used for this endeavor, such as Matlab®, Maple™, or Mathcad®. Mathematica® was selected for its concise language and strong symbolic and numeric capabilities. Also, it is available in an economic Home Edition, making self-study a practical option. The downside of Mathematica is that it can be somewhat cryptic. For those not familiar with the syntax, Appendix A decodes the expressions used in this book.

Mathematica will help us with much of the mathematical drudgery. However, there is no escaping the fact that PVD sputtering is a technical field. The target reader of this book will benefit most if they have completed the equivalent of a US undergraduate education in engineering or a physical science. This is because familiarity with calculus, differential equations and the basic physics of electricity and magnetism are necessary background for understanding the concepts discussed here. SI units will be used throughout, with a few exceptions for common units like torr and eV.

PVD sputtering is a huge field, and in this little book we will explore only a small portion of it. The path we will take is mostly that of an electron in the plasma. In the early chapters we will develop our understanding of the electric and magnetic forces that act upon it. We will then model the motion of the electron over the sputter target, seeing how the magnetic field traps it and how the combination of electric and magnetic fields keeps it and its brothers circulating near the target surface. Next, we will study how electron collisions produce the argon ions that do the actual sputtering. These collisions will then be incorporated into our model of electron motion to generate a model that can predict the erosion profile that is etched into the target. Finally, in the last chapter, we develop a method for predicting the deposition rate based on the erosion profile.

This book is written with the intent that you follow along on your computer, creating the models as we go. By doing this, you will develop a deeper understanding of the material and create a set of tools you can use in your own explorations of this fascinating topic. For those who want to save some time, Mathematica notebooks can be downloaded from www.ComputationalExplorations.com.

Authors have many reasons for writing. For me it is the desire to create and share a clear and useful reference for PVD modeling. In 2008, I joined a small company developing a new solar cell technology via PVD sputtering. Up until that time my work had primarily focused on chemical vapor deposition (CVD). Fortunately I had very knowledgeable colleagues and found many excellent papers on modeling various aspects of the PVD process.

Thank you to Dan Juliano, Rob Martinson and Paul Shufflebotham for furthering my understanding of PVD. I am also grateful to Fred Chetcuti for many stimulating discussions and to Kedar Hardikar, Sassan Roham, Pete Woytowitz, and Ron Powell for providing helpful feedback on this manuscript. Finally, I would like to thank my wife, Bonnie. Without her love and encouragement, this book would not have been possible.

San Jose, CA
September 2014

1. INTRODUCTION

1.1 WHAT IS PVD SPUTTERING?

PVD sputtering is a method for depositing thin films. Conceptually, a PVD system consists of just four components: the target being sputtered, a substrate receiving the sputtered material, a vacuum chamber, and a power supply. To deposit a thin film of say copper onto a substrate, we would place a copper block (the target) and the substrate into the vacuum chamber. After pumping the chamber down to a few millitorr, a small flow of argon would be used to maintain that pressure. We would then apply a large negative voltage to the copper block—such as -500 V, relative to the grounded chamber walls. The substrate could be grounded or floating. The electric field created by this voltage will induce a plasma to form inside the chamber and positive ions will be attracted to the copper. As they fall through the electric field, they strike the target with several hundred electron volts of energy. Since the copper atoms are bound together by only a few eV of energy, each ion has a good chance of knocking an atom off of the copper target, sending it into the vacuum. The liberated Cu atom flies across the chamber, hits the substrate, and sticks.

This process would quickly run out of ions without a mechanism to replenish them. Secondary electron emission is that mechanism. When the argon ion strikes the copper target, a secondary electron is sometimes emitted. This electron is accelerated away from the target by the same electric field that attracted the ion. It can pick up 500 eV of energy during this time. Since it only takes about 16 eV to ionize an argon atom, we see that a single secondary electron can create many ions. By this method, the plasma is

1

sustained and copper is deposited onto the plate, one sputtered atom at a time.

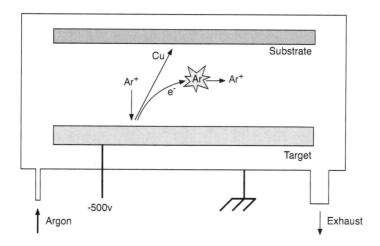

1.2 WHY PVD SPUTTERING?

There are many ways to deposit thin films, such as evaporation, electroplating, and chemical vapor deposition. Each has benefits and drawbacks. The primary benefits of PVD sputtering are

- High purity films. Because the films are sputtered in a high vacuum environment, there is very little contamination. If the target has high purity, the resulting film will also.
- Temperature flexibility. Sputtering can be done at room temperature allowing thin films to be deposited on delicate substrates, like plastics. It can also be done at high temperatures, if required, to achieve particular film properties.
- Compositional control. The deposited film typically has the composition of the target material.
- High deposition rates. By running at high power, deposition rates of 1 µm/min or higher can be achieved.
- Economical. In a well-designed PVD system, a large fraction of the target material is deposited onto the substrate.
- Environmentally friendly. Unlike plating or CVD, there are no hazardous chemicals used.

Because of these advantages, sputtered films are used in a great many products, spanning a variety of industries. In our computers, the hard drive and integrated circuits make heavy use of sputtered films. In fact, these films have played a key role in the staggering improvements we see in computer hardware each year. Our windows are coated with sputtered films to lower emissivity and boost energy efficiency. We pay extra to have antireflective coatings sputtered onto our eyeglasses to reduce glare. The potato chips we so enjoy are kept extra fresh due to a thin film of sputtered aluminum on the inside of the bag. Drill bits maintain their sharpness longer when sputtered with a thin layer of titanium nitride. And the lovely gold sheen on costume jewelry has been achieved with a sputtered coat of the same film.

1.3 MAGNETRON SPUTTERING

Today, nearly all sputtering is accomplished by magnetron sputtering. Developed in the 1970s, this technology uses a magnetic field to trap the electrons near the target surface. Since this technology will be the focus of our efforts, let's briefly describe some of its key aspects.

When a secondary electron is first liberated from the target surface, it is in a very uncomfortable position. It is sitting next to a large negative charge (our target is set to -500 V). This creates a very strong repulsive force on the electron and it quickly accelerates away from the target at high velocity. When traveling so quickly, the probability of hitting an argon atom and ionizing it is small. As a result, the electron may only ionize a few atoms before striking the chamber wall and being lost. This is an inefficient use of our electrons and energy.

The magnetron addresses this problem by trapping the electrons near the target surface. They then have the time to ionize many atoms before losing their energy.

In the PVD magnetron system, the electrons are subjected to two forces, one from the electric field present in the plasma and one from the magnetic field of the magnetron. These are summarized in the Lorentz force equation:

$$F = qE + qv \times B \qquad (1\text{-}1)$$

where q is the charge on the electron. If we look at just the magnetic portion, we see that it contains the cross product of velocity and the magnetic induction B. This indicates that the force is perpendicular to both the direction the electron is traveling and the local B field. This results in the electron spiraling around the magnetic flux lines. The radius of the electron spiral is called the Larmor radius or the gyroradius and is given by

$$r_L = \frac{mv_\perp}{eB} \qquad (1\text{-}2)$$

where m and e are the mass and charge of the electron, while $v_\perp$ is the component of velocity perpendicular to the B field. This ability to trap the electron around a flux line is a key aspect to the magnetron operation. This can be seen in Figure 1- 1, which shows an electron circling around a line.

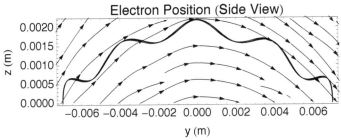

Figure 1- 1. Electron circling around a field line.

While the B field is key to trapping the electrons, it is the interaction between the electric and magnetic forces that makes this so successful. In the presence of both fields, the electrons experience a drift velocity:

$$v_{drift} = \frac{E \times B}{B^2} \qquad (1\text{-}3)$$

The direction of this drift is perpendicular to both fields. We can get a feel for the magnitude of this E cross B drift by considering some typical values. The -500 V applied to the cathode creates an E field in the plasma. The field is strongest near the target, in the *sheath* region. This region is typically a millimeter or so in thickness. The E field in this region is then approximately

$$E = \frac{500V}{0.001m} = 5 \times 10^5 \, V/m$$

The B field is typically in the 500 G to 1000 G range near the target surface. Let's pick the smaller value:

$$v_{drift} = \frac{E \times B}{|B|^2} = \frac{5 \times 10^5 \ V/m \times 0.05T}{(0.05\ T)^2} = 10^7 m/s$$

In order to get the velocity in meters per second, we converted the B field to SI units, noting that 10,000 gauss = 1 tesla. This speed indicates that even in a large magnetron, the electron completes a lap around the racetrack in less than a microsecond.

Magnetrons are designed so electrons experiencing this drift move in a closed path above the target surface. This keeps the electrons near the target, as their kinetic energy is consumed in collisions with argon atoms. A closed path for the electrons can be created by having a row of magnets of one polarity surrounded by magnets of the opposite polarity. The sputtering rate of the target is high beneath this path, and a track gets etched into the target here. It is referred to as the racetrack.

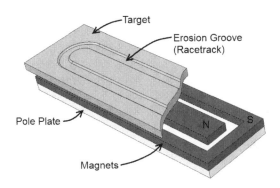

Figure 1- 2. Typical planar magnetron and target.

We can use the right-hand rule to determine the direction of electron travel. Noting the E field points positive to negative, and the B field points from north to south, the magnetron in Figure 1- 2 will cause electrons to drift clockwise.

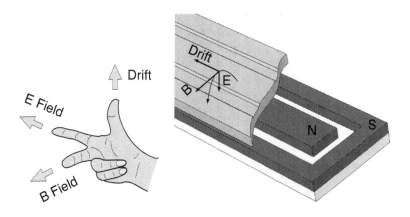

We can get a feel for the time scale of the magnetron by calculating how long it takes an electron to complete one loop. If our racetrack is two meters long, and the electron drifts at 10^7 m/s, we get

$$t = \frac{L}{v} = \frac{2m}{10^7 m/s} = 0.2\mu s$$

The other time scale of interest is the time needed for the electron to complete one orbit around a B field line. This is typically expressed in terms of the electron cyclotron frequency,

$$\omega_c = \frac{qB_0}{m}$$

For an electron, the electric charge is 1.602 x 10^{-19} C, and the mass is 9.11 x 10^{-31} kg.

$$\omega_c = \frac{(1.602 \times 10^{-19}C)(0.05T)}{9.11 \times 10^{-31}kg} = 8.79 \times 10^9 \text{ radians/sec}$$

$$t = \frac{2\pi}{\omega} = 7.15 \times 10^{-10} \text{ sec}$$

The electron will complete hundreds of cycles in the time required to go around the racetrack.

Sputtering tools come in all shapes and sizes. We will focus our efforts on a planar magnetron like in Figure 1- 2. The same principles apply to other configurations.

In the remainder of the book we will expand on these ideas in order to better understand the sputtering process. The first steps will be to flesh out our understanding of magnetic and electric fields. These fields can then be used to explore electron motion above the target surface. We will then look at electron collisions with gas atoms to better understand how ionization takes place. With that knowledge we can then predict the erosion profile etched into the target. Once the erosion profile is determined it can be used to predict deposition uniformity on the substrate.

.

2. MAGNETIC FIELD

What distinguishes magnetron sputtering from ordinary sputtering is, of course, the magnets. In Chapter 1 we saw how a magnetic field can trap an electron above the target surface. In this chapter we will further explore the magnetic field and develop an analytic expression for the B field of a single rectangular permanent magnet. This can then be used to calculate the field for an array of magnets on a pole plate.

2.1 MAGNETIC CHARGE APPROACH

It can be quite difficult to calculate the magnetic field generated by a magnetron. Yet we need this calculation in order to understand the electron motion above the target. What makes the problem difficult is the non-linear response of magnetic materials. For instance, the permeability of the steel pole plate can vary by 100x, depending on the local field. The permeability affects the field, so we end up in a situation where everything depends on everything else.

We can take a simplified approach, which will allow us to predict the magnetic field with quite good accuracy. (If greater accuracy is needed, a dedicated software package such as Maxwell (Ansys, Inc. 2014) or Lorentz (Integrated Engineering Software 2014) is recommended.) In a permanent magnet, the magnetic domains throughout the volume are pointing in the same direction—say, up. Our approach is to replace this volumetric magnetism with a surface charge on the top and bottom surfaces (the north and south faces). This charge is a magnetic charge. It doesn't exist in reality, but it is a useful device for our purposes. It allows us to simplify the problem and generate an expression for the magnetic field.

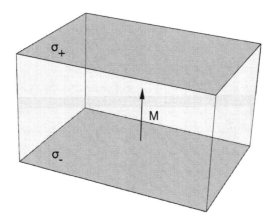

Figure 2- 1. The magnetization M can be replaced by equivalent surface "charges."

From Maxwell's equations we know that the curl of the magnetic field H is equal to the current density J:

$$\nabla \times H = J$$

This tells us the manner in which a current generates the field. In PVD sputtering, the currents generated inside the plasma are very weak in terms of the magnetic field they generate and they can be neglected. Then we can say

$$\nabla \times H = 0 \qquad (2\text{-}1)$$

From this, we can infer that H may be expressed in terms of the gradient of a potential:

$$H = -\nabla \phi_m \qquad (2\text{-}2)$$

(This comes about from the vector identity: $\nabla \times \nabla \phi = 0$, which says that when you take the curl of the gradient of any scalar function, the result is zero.) This is useful because it means we can look for a solution to ϕ_m, which a scalar, rather than a solution for H, which is a vector. We are actually interested in the B field. From another of Maxwell's equations, we know that the divergence of B is zero.

$$\nabla \cdot B = 0 \qquad (2\text{-}3)$$

The B field is related to the H field by

$$B = \mu H + \mu M \qquad (2\text{-}4)$$

where μ is the permeability and M is the magnetization of the material. Combining equations (2-3) and (2-4) we see

$$\nabla \cdot (\mu H + \mu M) = 0 \qquad (2\text{-}5)$$

$$\nabla \cdot H = -\nabla \cdot M \qquad (2\text{-}6)$$

In terms of the magnetic potential, this can be written as

$$\nabla \cdot (-\nabla \phi_m) = -\nabla \cdot M \qquad (2\text{-}7)$$

or

$$\nabla \cdot \nabla \phi_m = -\rho_m \qquad (2\text{-}8)$$

where we define $\rho_m = -\nabla \cdot M$. That is, we are defining the magnetic charge in terms of the magnetization of the material. The beauty of equation (2-8) is that it is in the form of Poisson's equation, which is commonly encountered in electrostatics and heat conduction. It is a much easier equation to solve than equation (2-1).

There is a general solution to equation (2-8), see Jackson (Jackson 1975), page 39. It is

$$\phi_m(x) = \frac{1}{4\pi} \int \frac{\rho_m}{|x - x'|} d^3 x' \qquad (2\text{-}9)$$

There are some key points to note. First, x is a position vector, specifically the x, y and z components of where ϕ is to be determined. We are integrating over a volume and this is the volume that contains the magnetic charges (i.e., the magnet). The position vector of a point in this volume is denoted by x'. So this is actually a triple integral, over x', y', and z'.

Our goal is to get an expression for B, not ϕ. To move in that direction, we take the gradient of both sides with respect to the observer position (x, not x'):

$$\nabla \phi_m(x) = \nabla \frac{1}{4\pi} \int \frac{\rho_m}{|x - x'|} d^3x' \qquad (2\text{-}10)$$

One can show (Jackson, page 33)

$$\frac{x - x'}{|x - x'|^3} = -\nabla \left(\frac{1}{|x - x'|} \right) \qquad (2\text{-}11)$$

Moving the gradient operator inside the integral in equation (2-10) and applying (2-2) and (2-11), we get

$$H(x) = \frac{1}{4\pi} \int \frac{\rho_m(x')\,(x - x')}{|x - x'|^3} d^3x' \qquad (2\text{-}12)$$

Recalling that in the absence of magnetic charges $B = \mu_0 H$, we find that

$$B(x) = \frac{\mu_0}{4\pi} \int \frac{\rho_m(x')\,(x - x')}{|x - x'|^3} d^3x' \qquad (2\text{-}13)$$

The last step is to replace ρ with $-\nabla \cdot M$. We use the Divergence theorem for this part:

$$\int \frac{\rho_m(x')}{|x - x'|^3} d^3x' = -\int \nabla \cdot M \, dx' = -\int M \cdot n \, da \qquad (2\text{-}14)$$

This tells us that the volumetric magnetization could be replaced with a magnetization on the surface of the magnets. Doing this would collapse our volume integrals to surface integrals. It will be helpful to keep it a volume integral, but note that M will be zero over the interior of the volume:

$$B(x) = \frac{\mu_0}{4\pi} \int \frac{M \cdot n\,(x - x')}{|x - x'|^3} d^3x' \qquad (2\text{-}15)$$

or

$$B(x) = \frac{1}{4\pi} \int \frac{B_r \cdot n\,(x - x')}{|x - x'|^3} d^3x' \qquad (2\text{-}16)$$

We are using the magnetic remanence, B_r, for the magnetization of our permanent magnet. This is the magnetization of the magnet after it has been removed from the charging unit. Table 1 shows the remanence for several types of magnets.

Table 1. Magnetic properties

Type	Max Energy Prod (MGOe)	Br (Tesla)
Nd-Fe-B	50	1.43
	45	1.36
	40	1.28
	35	1.2
	30	1.1
SM-Co	30	1.12
	26	1.05
	20	0.9

From Dexter Magnetic Technologies (Dexter Magnetic Technologies 2014)

Equation (2-16) can be used to calculate the B field of a single magnet or an array of magnets, provided we know B_r and the shape of the magnets.

2.2 THE B FIELD OF A SINGLE MAGNET—NUMERICALLY

Imagine we have a 50 MGOe Nb-Fe-B magnet from Table 1 and it is a cube of 1 cm x 1 cm x 1 cm. It is oriented so that its top face is north and bottom face is south. We can numerically integrate equation (2-16) to calculate the B field of this magnet. Using Mathematica's NIntegrate function, we can define functions for determining the magnetic field of this magnet anywhere in space. Here is the function for the x component of B:

Code 2- 1. Function for calculating Bx of a magnet

```
cm = 0.01;  (* conversion from cm to m *)

Bx[x_, y_, z_] := NIntegrate[
    Br (x - xm)
    -----------------------------------------------------  -
    4 π ((x - xm)² + (y - ym)² + (-h/2 + z)²)^(3/2)

    Br (x - xm)
    -----------------------------------------------------  , {xm, -0.5 cm, 0.5 cm},
    4 π ((x - xm)² + (y - ym)² + (h/2 + z)²)^(3/2)

    {ym, -0.5 cm, 0.5 cm}
]
```

The first line is simply a convenience, a conversion factor from centimeters to meters. This allows us to write out the expression in terms of centimeters, but all of the calculations are done in the MKSA unit system.

Because the magnetic charge is on the upper and lower surfaces of the magnet (i.e., at two discrete z values), we can solve this as two surface integrals over x and y. The sign difference between the two is due to one having positive magnetic charge while the other is negative.

Using this function, we can plot Bx above the magnet, as seen in Figure 2-2.

```
f22 = Plot[Bx[x, 0, 2 cm] * 10 000, {x, -5 cm, 5 cm},
    Frame -> True, Axes → None,
    BaseStyle → {Medium, FontFamily → "Helvetica"},
    FrameLabel -> {"x (m)", "Bx (G)"},
    PlotLabel → "Bx vs x", PlotStyle → Black,
    Prolog → {Gray,
      Rectangle[{-0.038, 60}, {-0.03, 90}],
      Black,
      Arrow[{{-0.044, 105}, {-0.024, 105}}]},
    Epilog → Line[{{-10, 0}, {10, 0}}],
    ImageSize → Large
  ]
```

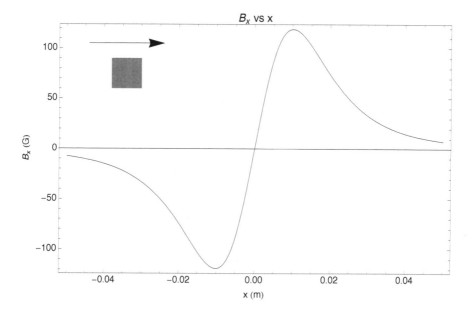

Figure 2- 2 **Bx for plotted along a line 2cm above a 1cm cube magnet with Br = 1.4T.**

2.3 ANALYTIC SOLUTION

The numerical integration of equation (2-16) is a perfectly legitimate and straightforward way to determine the B field of our magnetron. In practice, however, the calculation is slow. As we follow an electron over the target surface, we need to know the B field at hundreds or thousands of points. Numerically solving (2-16) at each point will slow our modeling to a crawl. Fortunately, there is an analytic solution to (2-16) that greatly speeds up the calculations.

We can use Mathematica to do the integral analytically, but it is quite tedious. First we must break up the integral into two surface integrals as we did for the numerical solution. We then integrate over x, then over y. Fortunately, Buyle and others have worked through this (G. Buyle 2005) (Engel-Herbert and Hesjedal 2005). Here is the solution:

$$B_x = \frac{|B_r|}{4\pi} \sum_{i,j,k=\pm 1} ijk \ln\left(Y + \sqrt{X^2 + Y^2 + Z^2}\right) \qquad (2\text{-}17)$$

$$B_y = \frac{|B_r|}{4\pi} \sum_{i,j,k=\pm 1} ijk \ln\left(X + \sqrt{X^2 + Y^2 + Z^2}\right) \qquad (2\text{-}18)$$

$$B_z = -\frac{|B_r|}{4\pi} \sum_{i,j,k=\pm 1} ijk \arctan\left(\frac{XY}{Z\sqrt{X^2 + Y^2 + Z^2}}\right) \qquad (2\text{-}19)$$

X, Y, and Z are the positions of the magnet surfaces; see Figure 2- 3. The magnet is assumed to be centered at the origin.

$$X = x + \frac{iW}{2} \quad Y = y + \frac{jL}{2} \quad Z = z + \frac{kH}{2}$$

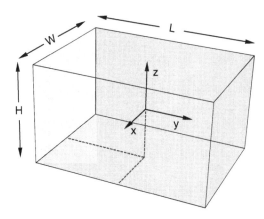

Figure 2- 3. Magnet coordinate system. The analytical solution assumes the magnet is centered at x = y = z = 0 and has dimensions H, W, and L.

It is a bit unclear, but these summations have $2^3 = 8$ terms in them. They sum over i, j and k, each of which can be -1 or 1.

In the above formulas the magnet is assumed to be centered at the origin and aligned with the coordinate system. We need to generalize this when we implement it in Mathematica.

A Mathematica function for Bx is given by

Code 2- 2. Function for Bx of a single magnet, based on equation (2-17).

```
BxO[{x_, y_, z_}, {{x0_, y0_, z0_}, {W_, L_, T_}, Br_}] := Module[{i, j, k, X, Y, Z},

    X = x - (x0 + (1 + i) W/2); Y = y - (y0 + (1 + j) L/2); Z = z - (z0 + (1 + k) T/2);

    -Br/4π

    (Sum[Sum[Sum[i j k Log[Y + √(X² + Y² + Z²)], {k, -1, 1, 2}], {j, -1, 1, 2}], {i, -1, 1, 2}])]]
```

This function is a bit more general than equation (2-17) in that the magnet can be in any general location. The lower left corner of the magnet is at coordinates (x0, y0, z0), and the magnet is still aligned with the coordinate system. We can make similar functions for By and Bz:

Code 2- 3. Functions for By and Bz of a single magnet based on Eqs (2-18) and (2-19).

```
By0[{x_, y_, z_}, {{x0_, y0_, z0_}, {W_, L_, T_}, Br_}] := Module[{i, j, k, X, Y, Z},

    X = x - (x0 + (1 + i) W/2); Y = y - (y0 + (1 + j) L/2); Z = z - (z0 + (1 + k) T/2);

    -Br
    ────
    4 π

    (Sum[Sum[Sum[i j k Log[X + √(X² + Y² + Z²) ], {k, -1, 1, 2}], {j, -1, 1, 2}], {i, -1, 1, 2}])]
```

```
Bz0[{x_, y_, z_}, {{x0_, y0_, z0_}, {W_, L_, T_}, Br_}] := Module[{i, j, k, X, Y, Z},

    X = x - (x0 + (1 + i) W/2); Y = y - (y0 + (1 + j) L/2); Z = z - (z0 + (1 + k) T/2);

    Br
    ────
    4 π

    (Sum[Sum[Sum[i j k ArcTan[ X Y / (Z √(X² + Y² + Z²)) ], {k, -1, 1, 2}], {j, -1, 1, 2}], {i, -1, 1, 2}])]
```

With these functions, we can calculate the B field anywhere in space (except where the denominator in equation (2-19) goes to zero!). Figure 2- 4 shows a vector plot of the field lines around our 1 cm target. It took twenty seconds to generate this plot on my 2012 MacBook Air computer. For comparison, generating the same plot using numerical integration takes 305 seconds, about fifteen times longer.

```
(vp2 = StreamPlot[{
    By[0, y, z],
    Bz[0, y, z]},
    {y, -5 cm, 5 cm},
    {z, -5 cm, 5 cm},
    AspectRatio → 5 / 5, StreamStyle → Black,
    BaseStyle → {Medium, FontFamily → "Helvetica"},
    PlotLabel → "B Field Vectors", FrameLabel → {"y (m)", "z (m)"},
    Epilog → {Gray, Rectangle[{-0.5 cm, -0.5 cm}, {0.5 cm, 0.5 cm}]},
    ImageSize → Large]) // Timing
```

B Field Vectors

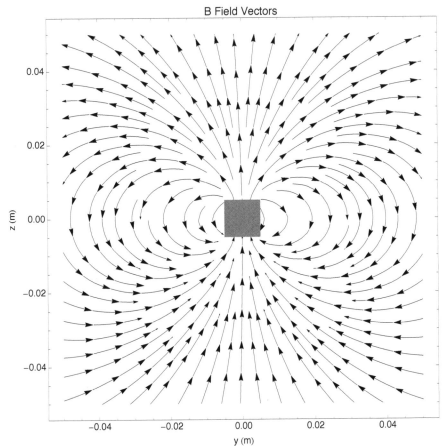

Figure 2- 4. Magnetic Field vectors. B field calculated using equations (2-18) and (2-19).

2.4 ARRAYS OF MAGNETS

The simplest magnet arrays we can generate are simple rows of magnets aligned with the coordinate system.

We can represent our array in Mathematica as a list, with each element of the list containing three items: the position of the bottom corner of the magnet, the size of the magnet, and the magnet's strength. The format is as follows:

```
{{x0, y0, z0}, {L, W, H}, Br}
```

Here is a list for a simple array consisting of magnets 1 cm wide, 1 cm tall and of varying lengths.

```
cm = 0.010; Br = 1.4;
magpack = {
   {{-17.5 cm, -1 cm, 0}, {35 cm, 2 cm, 1 cm}, Br}, (*center mags*)
   {{-20 cm, -2.5 cm, 0}, {1 cm, 5 cm, 1 cm}, -Br}, (*bottom row*)
   {{-20 cm, 2.5 cm, 0}, {40 cm, 1 cm, 1 cm}, -Br}, (*right side*)
   {{19 cm, -2.5 cm, 0}, {1 cm, 5 cm, 1 cm}, -Br}, (*top row*)
   {{-20 cm, -3.5 cm, 0}, {40 cm, 1 cm, 1 cm}, -Br}} (*left side*)
```

To visualize this we can use the Graphics3D command. Graphics3D will draw simple 3-D shapes for us, like blocks and spheres. To draw a block, we follow the Graphics3D command with the Cuboid command, and specify the lower left and upper right corners of the block. I use the Map function to convert the magpack coordinates into the lower left and upper right coordinates needed by Cuboid.

```
g1 = Graphics3D[{Gray, Opacity[1], Map[Cuboid[#[[1]], #[[1]] + #[[2]]] &, magpack]},
   BaseStyle → {Medium, FontFamily → "Helvetica"},
   Boxed → False,
   Axes → True,
   AxesLabel → {"x (m)", "y (m)"},
   ImageSize → Large, Lighting → "Neutral"]
```

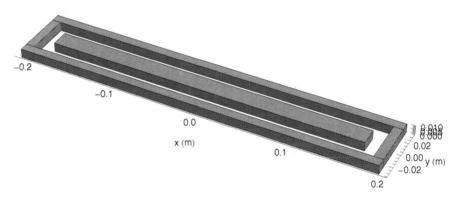

Figure 2- 5. Array of magnets.

We can use our single magnet equations to process all of the magnets in our array. We use the Map function to calculate the field of each magnet, and then we sum them up using Total. One of our assumptions is that the solution is linear—that is, the B field at any given point is the sum of the fields from all magnets. While not strictly true, we can get very accurate results with this assumption.

Code 2- 4. Functions for the magnetic field of an array of magnets.

```
Clear[Bx, By, Bz]
```

```
Bx[{x_, y_, z_}, mags_] :=
  Total[
    Map[BxO[{x, y, z}, #] &, mags]]
```

```
By[{x_, y_, z_}, mags_] :=
  Total[
    Map[ByO[{x, y, z}, #] &, mags]]
```

```
Bz[{x_, y_, z_}, mags_] :=
  Total[
    Map[BzO[{x, y, z}, #] &, mags]]
```

Using these functions, we can plot the B field in a line across the width of the magpack:

```
Plot[By[{0 cm, y, 2 cm}, magpack] * 10000, {y, -6.5 cm, 6.5 cm}, Frame -> True, Axes -> None,
  BaseStyle -> {Medium, FontFamily -> "Helvetica"}, FrameLabel -> {"y (m)", "B_y (G)"},
  PlotStyle -> Black, PlotLabel -> "B_y vs y",
  Prolog -> {Gray,
    Rectangle[{0.02, -1400}, {0.026, -1100}],
    Rectangle[{0.032, -1400}, {0.044, -1100}],
    Rectangle[{0.050, -1400}, {0.056, -1100}],
    Black,
    Arrow[{{0.02, -900}, {0.056, -900}}]},
  Epilog -> Line[{{-10, 0}, {10, 0}}],
  ImageSize -> Large]
```

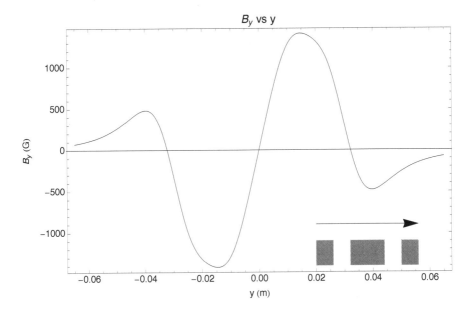

Figure 2- 6. B_y in a slice across the magnet array

We can see how the individual magnets combine to give this profile:

```
Plot[{
  By0[{0 cm, y, 2 cm}, magpack[[1]]] * 10 000,
  By0[{0 cm, y, 2 cm}, magpack[[3]]] * 10 000,
  By0[{0 cm, y, 2 cm}, magpack[[5]]] * 10 000},
  {y, -6.5 cm, 6.5 cm},
  Frame -> True, Axes → None,
  BaseStyle → {Medium, FontFamily → "Helvetica"},
  PlotStyle → {Black, {Black, Dashed}, {Black, Dashed}},
  FrameLabel -> {"y (m)", "B_y (G)"},
  PlotLabel → "B_y vs y",
  Prolog → {Gray,
    Rectangle[{0.025, 750}, {0.031, 910}],
    Rectangle[{0.037, 750}, {0.049, 910}],
    Rectangle[{0.055, 750}, {0.061, 910}],
    Black,
    Arrow[{{0.025, 1000}, {0.061, 1000}}]},
  Epilog → Line[{{-10, 0}, {10, 0}}],
  ImageSize → Large]
```

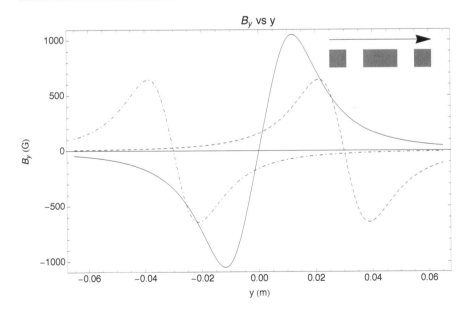

Figure 2- 7. B_y for each magnet in array.

The racetrack, which is the path the electrons follow over the target surface, is defined by Bz = 0. We can see where this is located by plotting Bz for a slice through the center of the magpack (Figure 2- 8).

```
Plot[Bz[{0 cm, y, 2 cm}, magpack] * 10 000, {y, -6.5 cm, 6.5 cm}, Frame -> True, Axes -> None,
  BaseStyle -> {Medium, FontFamily -> "Helvetica"},
  PlotStyle -> Black, FrameLabel -> {"y (m)", "Bz (G)"},
  PlotLabel -> "Bz vs y",
  Prolog -> {Gray,
    Rectangle[{0.02, 1200}, {0.026, 1400}],
    Rectangle[{0.032, 1200}, {0.044, 1400}],
    Rectangle[{0.050, 1200}, {0.056, 1400}],
    Black,
    Arrow[{{0.02, 1550}, {0.056, 1550}}]},
  Epilog -> {Line[{{-10, 0}, {10, 0}}],
    Arrow[{{-0.037, 900}, {-0.018, 50}}],
    Text["Bz=0", {-.04, 1000}]},
  ImageSize -> Large]
```

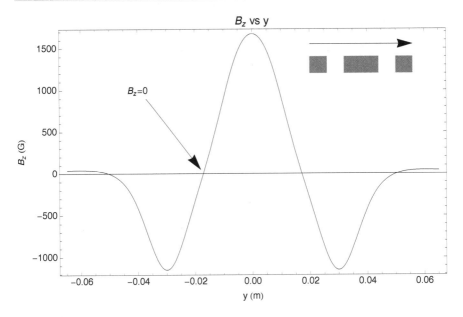

Figure 2- 8. Bz in slice across magnet array.

A better way to visualize the racetrack is to plot it on the target surface. First we will plot contours of the B field parallel to the target surface. Then we will overlay the racetrack on top. This parallel component is of interest because it, combined with the E field normal to the surface, gives us the ExB drift that moves the electrons around the racetrack.

We can create a small function to determine the parallel B field using our Bx and By functions. The Norm function calculates the magnitude of the vector in the x-y plane.

```
Bpara[{x_, y_, z_}, mags_] :=
  Norm[{Bx[{x, y, z}, mags], By[{x, y, z}, mags]}]
```

```
Bpara[{1 cm, 1 cm, 2 cm}, magpack]
```

```
0.128328
```

```
cp1 = ContourPlot[
  Bpara[{x , y , 2 cm}, magpack],
  {x, -20 cm, 20 cm}, {y, -3.5 cm, 3.5 cm},
  AspectRatio → 3.5 / 20, ColorFunction → ColorData["GrayTones"], ImageSize → Large]
```

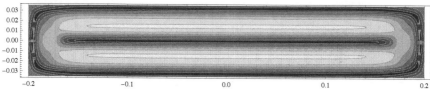

Figure 2- 9. Contours of B field parallel to target surface.

Zooming in on the turnaround region on the right:

```
cp1 = ContourPlot[
  Bpara[{x , y , 2 cm}, magpack],
  {x, 10 cm, 20 cm}, {y, -3.5 cm, 3.5 cm},
  AspectRatio → 3.5 / 5, ColorFunction → ColorData["GrayTones"],
  ImageSize → Large]
```

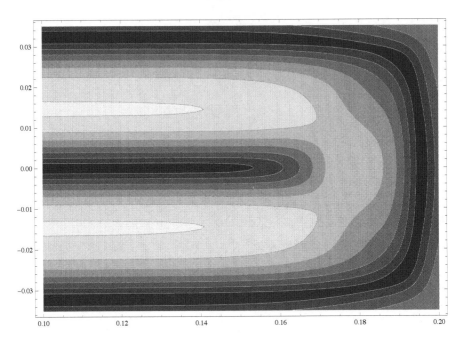

Figure 2- 10. Component of B field parallel to target.

This plot shows us that the B field is weaker at the ends compared to the middle. From equation (1- 3), a weaker field means a higher drift velocity, which has consequences for target utilization.

The racetrack is also easily generated by plotting the contour Bz = 0:

```
cp2 = ContourPlot[
  Bz[{x, y, 2 cm}, magpack] == 0.,
  {x, -20 cm, 20 cm}, {y, -3.5 cm, 3.5 cm},
  AspectRatio → 3.5 / 20, ContourStyle → {Black, Thick},
  ImageSize → Large]
```

Figure 2- 11. The racetrack is defined by Bz = 0

This can be overlaid on top of the Bpara plot using the Show command:

24

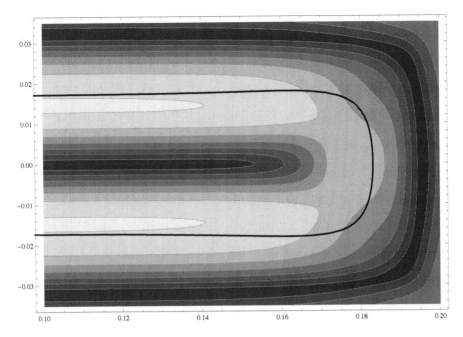

Figure 2- 12. Racetrack overlaid on top of Bpara contour plot

This overlay shows an interesting result. The racetrack, where Bz = 0, does not correspond to where Bpara is maximum. In general, the racetrack is slightly outboard of the peak in Bpara values. This comes about because both the magnitude of the field as well as its direction are varying over the target surface.

Another useful way to visualize the field of our magnetron is to plot the field lines. We can use the StreamPlot function in Mathematica to do this. To look at B field vectors projected onto the target surface, we can plot the Bx and Bz components of B in the plane of the target.

```
vp1 = StreamPlot[
    {Bx[{x, y, 2 cm}, magpack], By[{x, y, 2 cm}, magpack]}, {x, -20 cm, 20 cm},
    {y, -3.5 cm, 3.5 cm}, AspectRatio → 3.5 / 20,
    BaseStyle → {Medium, FontFamily → "Helvetica"}, StreamStyle → Black,
    PlotLabel → "B Field Vectors", FrameLabel → {"x (m)", "y (m)"},
    Prolog → Prepend[
        Map[Rectangle[#[[1, 1 ;; 2]], #[[1, 1 ;; 2]] + #[[2, 1 ;; 2]]] &, magpack], LightGray],
    ImageSize → Large]
```

2. Magnetic field

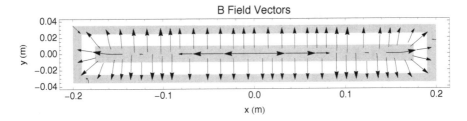

Figure 2- 13. Magnetic field 2 cm above magnet array.

Zooming in on the right end, we can see more detail, especially if we overlay the racetrack path.

```
vp2 =
StreamPlot[{Bx[{x, y, 2 cm}, magpack], By[{x, y, 2 cm}, magpack]}, {x, 10 cm, 20 cm},
    {y, -3.5 cm, 3.5 cm}, AspectRatio → 3.5 / 5,
    BaseStyle → {Medium, FontFamily → "Helvetica"}, StreamStyle → Black,
    PlotLabel → "B Field Vectors", FrameLabel → {"x (m)", "y (m)"},
    Prolog → Prepend [
        Map[Rectangle[#〚1, 1 ;; 2〛, #〚1, 1 ;; 2〛 + #〚2, 1 ;; 2〛] &, magpack], LightGray],
    ImageSize → Large]
```

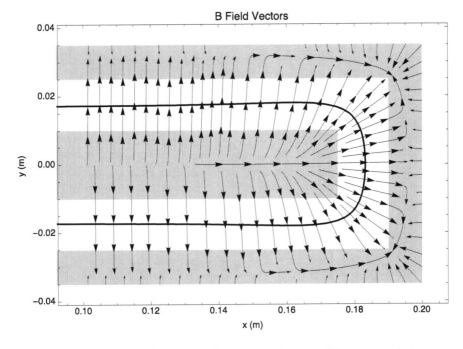

Figure 2- 14. Magnetic field 2 cm above magnet array. The racetrack is superimposed.

In these plots we have drawn in the magnet locations (shaded regions) using the Prolog option to StreamPlot. We can see that for this magnetron, the racetrack is close to halfway between the magnets. If the inner and outer magnets were of different strengths, or oriented differently, this would not be true.

We can use these B field functions to see how the racetrack shape changes with distance above the magnet array. For instance, as we erode into the target, does the racetrack widen or narrow? The code below finds where Bz = 0 for a range of heights above the magpack and plots the result.

```
ListPlot[{Map[{ y /. FindRoot[Bz[{0 cm, y, #}, magpack], {y, -2.5 cm}], #} &,
    Range[1.4 cm, 4.5 cm, 0.1 cm]],
  Map[{ y /. FindRoot[Bz[{0 cm, y, #}, magpack], {y, 2.5 cm}], #} &,
    Range[1.4 cm, 4.5 cm, 0.1 cm]]},
  Frame -> True, Axes -> None,
  BaseStyle -> {Medium, FontFamily -> "Helvetica"},
  FrameLabel -> {"y (m)", "z (m)"}, PlotStyle -> Black,
  PlotLabel -> "Racetrack Position vs Height",
  PlotRange -> {{-4 cm, 4 cm}, {0, 5 cm}},
  Epilog -> Prepend [Map[Rectangle[#[[1, 2 ;; 3]], #[[1, 2 ;; 3]] + #[[2, 2 ;; 3]]] &,
    {magpack[[1]], magpack[[3]], magpack[[5]]}], LightGray],
  Prolog -> {Text["Racetrack", {0, 0.03}], Arrow[{{-0.005, 0.03}, {-0.017, 0.03}}]],
    Arrow[{{0.005, 0.03}, {0.017, 0.03}}]]},
  ImageSize -> Large]
```

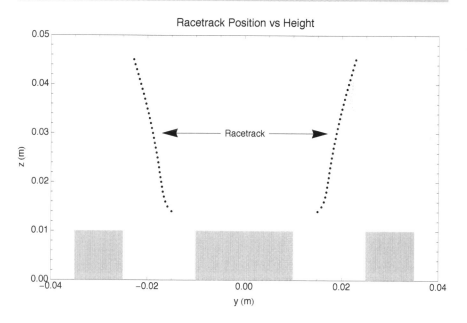

Figure 2- 15. The width of the racetrack increases with distance from the magnet array.

For this magnet array, the racetrack broadens when moving away from the magnets. We can use this same method to see how quickly the B field drops off with distance from the magnets. We calculate Bnorm at the Bz = 0 points as we move away from the magnets. The figure below shows this. The field drops off quickly with distance, falling from about 2000 G at 1.5 cm to about 300 G at 3.5 cm. This fast drop-off limits how thick a target can practically be.

```
ListLinePlot[Map[{100 #[[1]], 10⁴ Bpara[{0, #[[2]], #[[1]]}, magpack]} &, zeroPts],
   Frame -> True, Axes → None,
   BaseStyle → {Medium, FontFamily → "Helvetica"},
   FrameLabel -> {"z (cm)", "Bₚₐᵣₐ (G)"}, PlotStyle → Black,
   PlotLabel → "Bₚₐᵣₐ vs z", ImageSize → Large]
```

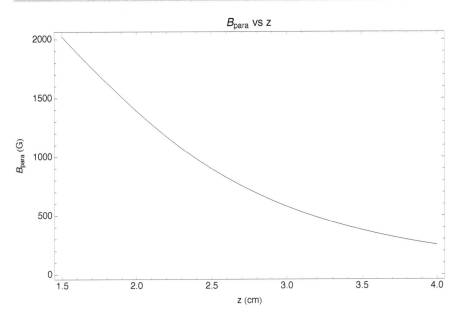

Figure 2- 16. The B field drops rapidly with vertical distance from the magnet array.

2.5 POLE PLATES

Normally the magnets in our magnetron are sitting on an iron plate. This provides an easy return path for the field lines, resulting in a stronger field above the magnets. So far we have completely ignored this plate, except to say it makes calculating the field difficult. However, we need to include it in some fashion, since it is present in virtually all magnetrons.

Following Buyle (G. Buyle 2005), we can include it by using the Method of Images. This is a technique from electrostatics for solving problems involving charged particles in the vicinity of a conductor. The conductor is at constant potential, typically grounded. If we can replace the conductor with an imaginary charge that gives the same constant potential at the boundary, we will have turned a difficult problem into a simple one. Figure 2- 17 shows the idea.

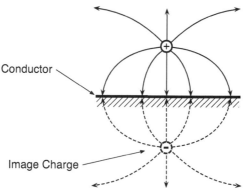

Figure 2- 17. Method of Images for charges.

In a similar way a high-permeability pole plate can be thought of as a good conductor of magnetic charge; thus, it maintains a constant magnetic potential. We can then replace the pole plate with a mirror set of magnetic charges. When the magnet is sitting on the pole plate, this is equivalent to doubling the height of the magnet (Figure 2- 18). This means the pole plate can be included by simply doubling the height of the magnets.

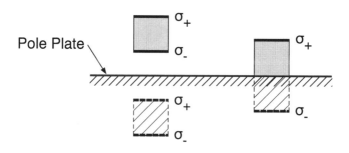

Figure 2- 18. Applying Method of Images to the pole plate.

Let's try out the method for our magpack. To simulate the pole plate, we will double the height of each magnet in our definition:

Code 2- 5. Definition of magnet array

```
magpack2 = {
  {{-17.5 cm, -1 cm, -1 cm}, {35 cm, 2 cm, 2 cm}, Br}, (*center mags*)
  {{-20 cm, -2.5 cm, -1 cm}, {1 cm, 5 cm, 2 cm}, -Br}, (*bottom row*)
  {{-20 cm, 2.5 cm, -1 cm}, {40 cm, 1 cm, 2 cm}, -Br}, (*right side*)
  {{19 cm, -2.5 cm, -1 cm}, {1 cm, 5 cm, 2 cm}, -Br}, (*top row*)
  {{-20 cm, -3.5 cm, -1 cm}, {40 cm, 1 cm, 2 cm}, -Br}}(*left side*)
```

We can now compare our nonpole piece magnetron to the one with a pole piece.

```
Plot[{By[{0 cm, y, 2 cm}, magpack] * 10 000,
  By[{0 cm, y, 2 cm}, magpack2] * 10 000},
  {y, -6.5 cm, 6.5 cm}, Frame -> True, Axes → None,
  BaseStyle → {Medium, FontFamily → "Helvetica"}, FrameLabel -> {"y (m)", "B_y (G)"},
  PlotLabel → "B_y vs y",
  PlotStyle → {Black, {Black, Dashed}},
  Prolog → {Gray,
    Rectangle[{0.02, -1700}, {0.026, -1400}],
    Rectangle[{0.032, -1700}, {0.044, -1400}],
    Rectangle[{0.050, -1700}, {0.056, -1400}],
    Black,
    Arrow[{{0.02, -1100}, {0.056, -1100}}]},
  Epilog → {Line[{{-10, 0}, {10, 0}}],
    Black,
    Line[{{-0.06, 1500}, {-0.055, 1500}}],
    Black, Dashed,
    Line[{{-0.06, 1100}, {-0.055, 1100}}],
    Text["No Pole Piece", {-0.043, 1500}],
    Text["Pole Piece", {-0.046, 1100}]},
  ImageSize → Large]
```

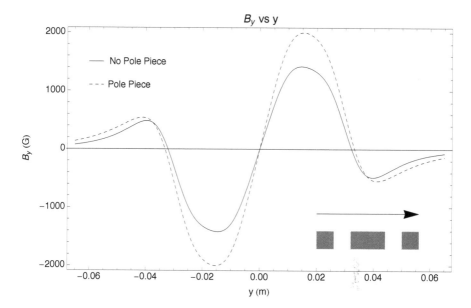

Figure 2- 19. B field with and without pole plate. Field calculated 2 cm above magnet array.

As expected, adding the pole piece boosts the strength of the magnetic field at the target surface.

3. ELECTRIC FIELD

Most of the plasma is electrically neutral and has little or no electric field. However, the story is different near the walls where large E fields can exist. At the target surface, in particular, these fields play an important role in the sputtering process. The ions are accelerated by the E field at the target, gaining enough energy to sputter atoms off the target. Then any secondary electrons released are accelerated by this same field, gaining the energy they need to ionize neutrals and sustain the plasma.

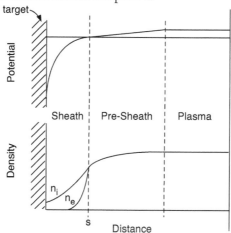

Figure 3- 1. A large electric field forms near the target surface.

These E fields near the walls come about in two ways. First, any surface in a plasma will serve as a recombination site for ions and electrons. This will partially deplete species near the wall region. Due to their high mobility, the electrons are more depleted than the ions, setting up a charge imbalance that creates the electric field. The region over which this imbalance occurs is called the plasma sheath. The voltage drop across such a sheath is typically around 20 V.

The second way an E field is created is through the application of a voltage directly to the wall surface. In PVD sputtering a voltage of -200 V or more is applied. This potential repels electrons from the wall region and attracts ions. Very large electric fields can be produced this way.

The ion and electron concentrations near a wall are sketched in Figure 3-1. It shows the plasma divided into three regions. In the sheath, as mentioned, the ion concentration exceeds the electron concentration. In the bulk plasma, the two are equal. There is a transition region connecting these called the pre-sheath. This region has a small electric field that provides some modest acceleration of ions and electrons.

The sheath is the key to understanding the E field in the plasma. For that reason, we explore various sheath models in this chapter.

3.1 ION MATRIX SHEATH

The simplest sheath model is the Ion Matrix Sheath. In this model we assume the sheath region has lost all of its electrons and the remaining ions are uniformly dispersed. This can occur when a field is first applied to an electrode: the electrons fly to the wall before the ions have time to react.

The potential in the sheath can be found from Poisson's equation

$$\nabla^2 \phi = -\frac{e n_i}{\epsilon_0} \tag{3-1}$$

where ϕ is the potential and e is the elementary charge. Because ion density, n_i, is a constant and the sheath is one-dimensional, we can easily integrate this once.

$$\frac{d\phi}{dx} = -\frac{e n_i}{\epsilon_0} x + C \tag{3-2}$$

Noting that $E = -\nabla \phi$, we can rewrite this as

$$E = \frac{e n_i}{\epsilon_0} x - C \tag{3-3}$$

We know the E field goes to approximately zero at $x = s$, the sheath boundary. Applying this boundary condition, we can solve for C

$$C = \frac{en_i}{\epsilon_0} s \tag{3-4}$$

Our expression for the electric field becomes

$$E = -\frac{en_i}{\epsilon_0}(s - x) \tag{3-5}$$

Figure 3-2 shows what the E field looks like in the sheath for two typical plasma densities of 10^{16} m^{-3} and 10^{17} m^{-3}. This is assuming a sheath thickness of 1 mm.

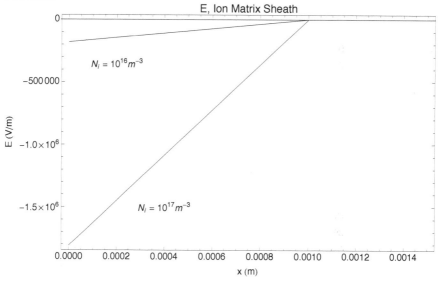

Figure 3- 2. Electric field in the ion matrix sheath for two different plasma densities.

One more integration gives us the potential ϕ.

$$\phi = -\frac{en_i}{2\epsilon_0}(s - x)^2 + C$$

$$\phi = \frac{en_i}{2\epsilon_0}x^2 - \frac{en_i}{\epsilon_0}sx + C \tag{3-6}$$

By setting ϕ to be zero at the sheath boundary, we find the integration constant to be $\frac{en_i}{2\epsilon_0}s^2$. The final equation is

$$\phi = -\frac{en_i}{2\epsilon_0}(s-x)^2 \tag{3-7}$$

A plot of ϕ is shown in Figure 3- 3 for the same two plasma densities.

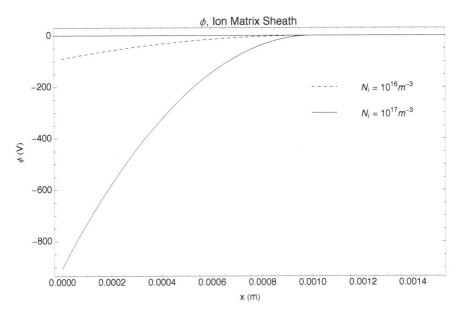

Figure 3- 3. **Potential in ion matrix sheath for two plasma densities.**

If we say the voltage on the target surface is $\phi = -\phi_w$, then we can solve for the sheath thickness:

$$s = \sqrt{\frac{2\epsilon_0\phi_w}{e\,n_i}} \tag{3-8}$$

This shows that the sheath thickness will increase as the voltage increases, while higher ion densities lead to thinner sheaths. For a plasma density of 10^{16} m^{-3}, and 500 V applied to the target, the sheath would be predicted to be

$$s = \sqrt{\frac{2\,(8.85\times10^{-12})\,500}{(1.602\times10^{-19})10^{16}}} = 0.74mm$$

The linear E field of the ion matrix sheath is the one most commonly used in Monte Carlo erosion models, and we will use it in Chapter 4. However, we made some pretty big assumptions in creating this sheath model. In the next sections we will explore some alternative approaches. It is worth noting that experimental measurements do show that the electric field is generally linear in the magnetron sheath (Choi, Bowden and Muraoka 1996).

3.2 CHILD'S LAW SHEATH

The next level of refinement in sheath modeling is to enforce conservation of ions. As the ions are attracted to the target, they pick up speed. Because we are assuming no ionization in the sheath, the flux of ions entering the sheath must be the same as the flux reaching the target. Thus,

$$n_0 v_0 = n\,v \qquad (3\text{-}9)$$

where the 0 subscript refers to conditions at the sheath boundary. As the ion velocity increases, the density must decrease. If we assume the ions enter the sheath with near zero velocity, then their energy is a simple function of position

$$\frac{1}{2} M\, v^2 = e\phi(s) - e\phi(x) \qquad (3\text{-}10)$$

where M is the ion mass and $e\phi(x)$ is the potential energy of the ion at position x. At the sheath boundary, where $x = s$, ϕ is zero and the ions have no kinetic energy. As the ions move into the sheath, they encounter a negative potential and pick up speed and energy. It will prove awkward to carry the s-x term everywhere. For this derivation it is more natural to have our coordinate system run from the sheath edge to the target, so we define a coordinate u:

$$u = s - x \qquad (3\text{-}11)$$

3. Electric field

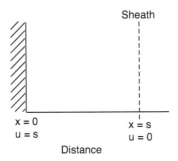

Sheath

x = 0
u = s

x = s
u = 0

Distance

Solving for ion velocity, we see

$$v = \sqrt{\frac{-2e\phi(u)}{M}} \qquad (3\text{-}12)$$

We have chosen the positive root because the velocity is positive in our new coordinate system. If we note that the current to the target surface is $j_i = en_0v_0$, then we can write out the ion density as

$$n_i(u) = \frac{j_i}{e\sqrt{\frac{-2e\phi(u)}{M}}} \qquad (3\text{-}13)$$

Since we are still assuming no electrons present in the sheath, Poisson's equation becomes

$$\frac{d^2\phi}{du^2} = -\frac{1}{\epsilon_0}\frac{j_i}{\sqrt{\frac{2e}{M}}}(-\phi)^{-1/2} \qquad (3\text{-}14)$$

This can be integrated once if we recall that $\left(\left(\frac{d\phi}{du}\right)^2\right)' = 2\frac{d\phi}{du}\frac{d^2\phi}{du^2}$, by the chain rule. We multiply both sides of the above equation by $2d\phi/du$ and integrate:

$$\int 2\frac{d\phi}{du}\frac{d^2\phi}{du^2}dx = \int \frac{e}{\epsilon_0}\frac{j_i}{\sqrt{\frac{2e}{M}}}(-\phi)^{-1/2}\left(2\frac{d\phi}{du}\right)du$$

$$\left(\frac{d\phi}{du}\right)^2 = \frac{2j_i}{\epsilon_0}\sqrt{\frac{M}{2e}}\int(-\phi)^{-\frac{1}{2}}d\phi$$

$$\left(\frac{d\phi}{du}\right)^2 = \frac{2J_i}{\epsilon_0}\sqrt{\frac{2M}{e}}(-\phi)^{\frac{1}{2}} \tag{3-15}$$

Next we take the square root of both sides and integrate

$$\int(-\phi)^{-\frac{1}{4}}d\phi = \sqrt{\frac{2J_i}{\epsilon_0}\sqrt{\frac{2M}{e}}}\int du$$

$$(-\phi)^{\frac{3}{4}} = \frac{3}{2}\left(\frac{J_i}{\epsilon_0}\right)^{\frac{1}{2}}\left(\frac{2M}{e}\right)^{\frac{1}{4}}u$$

We can now switch back to our normal coordinate system by replacing u with s-x:

$$(-\phi)^{\frac{3}{4}} = \frac{3}{2}\left(\frac{J_i}{\epsilon_0}\right)^{\frac{1}{2}}\left(\frac{2M}{e}\right)^{\frac{1}{4}}(s-x) \tag{3-16}$$

At the target surface, $x = 0$ and the potential is ϕ_w. Substituting these values we can determine the sheath thickness

$$s = \frac{2}{3}\left(\frac{J_i}{\epsilon_0}\right)^{-\frac{1}{2}}\left(\frac{2m}{e}\right)^{-\frac{1}{4}}\phi_w^{\frac{3}{4}} \tag{3-17}$$

This is the famous Child's law, relating the sheath thickness to the current and wall potential. This shows the sheath thickness increasing with potential and decreasing with current.

We can plot this expression to better see these relationships. Let's assume that our erosion racetrack is on average 2 cm wide and the length of the racetrack is 2 m. The eroded area is then 0.04 m². We can calculate the current density by picking a current and dividing by this area. Below we look at three currents: 5, 10, and 20 A.

```
mAr = 40 / (6.022 × 10²⁶)   (*  mass of argon atom, kg  *)
```

```
6.64231 × 10⁻²⁶
```

```
sChild[j_, ϕ_] := Module[{ε0 = 8.85 × 10⁻¹², e = 1.602 × 10⁻¹⁹},
  2/3 (j/ε0)^(-1/2) (2 e/mAr)^(1/4) ϕ^(3/4)]
```

```
Plot[{sChild[5 / 0.04, φ] * 1000,
  sChild[10 / 0.04, φ] * 1000, sChild[20 / 0.04, φ] * 1000},
 {φ, 200, 700}, Frame -> True, Axes → None,
 BaseStyle → {Medium, FontFamily → "Helvetica"},
 FrameLabel -> {"Φ (V)", "Sheath Thickness (mm)"},
 PlotLabel → "Sheath Thickness vs Potential",
 PlotStyle → Black,
 PlotRange → All,
 Epilog → {
   Text["5A", {340, 0.7}],
   Text["10A", {375, 0.54}],
   Text["20A", {400, 0.4}]},
 ImageSize → Large]
```

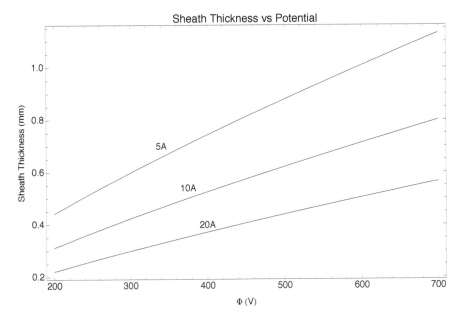

Figure 3- 4. Sheath thickness versus wall potential and current.

For the conditions we picked, the sheath varies between 0.2 and 1 mm. At 500 V and 10 A, it is approximately 0.6 mm thick.

We can use equation (3-16) to see how the potential varies in the sheath. The Mathematica Solve command can solve for ϕ and then generate a function to be plotted. Assume a 1 mm thick sheath.

$$\text{Solve}\left[(-\phi[u])^{3/4} = \frac{3}{4}\, 2^{3/4}\, u \sqrt{\frac{j i \sqrt{\frac{M}{e}}}{\epsilon 0}}\, ,\, \phi[u] \right] \,/.\, u \to s - x$$

$$\left\{\left\{\phi[s-x] \to -\frac{3 \times 3^{1/3}\,(s-x)\left((s-x)\sqrt{\frac{j i \sqrt{\frac{M}{e}}}{\epsilon 0}}\right)^{1/3}\sqrt{\frac{j i \sqrt{\frac{M}{e}}}{\epsilon 0}}}{2 \times 2^{2/3}}\right\}\right\}$$

$$\text{phiChild}[x\_] := -\frac{3 \times 3^{1/3}\,(s-x)\left((s-x)\sqrt{\frac{j i \sqrt{\frac{M}{e}}}{\epsilon 0}}\right)^{1/3}\sqrt{\frac{j i \sqrt{\frac{M}{e}}}{\epsilon 0}}}{2 \times 2^{2/3}}$$

```
Plot[{phiChild[x] /. consts2 /. ji → 5 / 0.04 /. s → 0.001,
  phiChild[x] /. consts2 /. ji → 10 / 0.04 /. s → 0.001,
  phiChild[x] /. consts2 /. ji → 15 / 0.04 /. s → 0.001
}, {x, 0, .0015},
Frame -> True, Axes → None,
BaseStyle → {Medium, FontFamily → "Helvetica"},
FrameLabel -> {"x (m)", "φ (V)"},
PlotLabel → "φ, Child's Law Sheath",
PlotStyle → {Black},
PlotRange → All,
Epilog → {Line[{{-10, 0}, {10, 0}}],
  Line[{{0.0010, -300}, {0.0011, -300}}],
  Text["5A", {0.0002, -400}],
  Text["10A", {0.0002, -650}],
  Text["20A", {0.0002, -850}]},
ImageSize → Large]
```

```
consts2 = {e->1.602x10⁻¹⁹, ε₀->8.85x10⁻¹², M->40/(6.022x10²⁶)};
```

3. Electric field

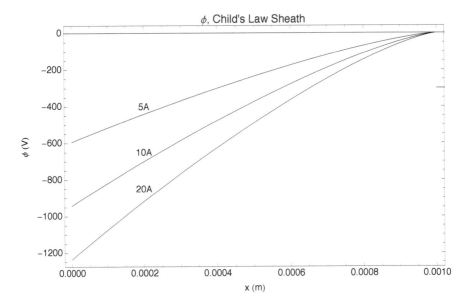

Figure 3- 5. Potential in the sheath for three currents

From Figure 3- 5, we can see that higher wall potentials are needed to drive higher currents through the sheath.

We can plot the E field by taking the derivative of the potential solution. This is done in Figure 3- 6. We see that the shape of the curve is quite different from the Ion Matrix model. The field is flatter near the wall, but drops off more steeply near the sheath boundary.

```
Plot[{D[-phiChild[x], x] /. consts2 /. ji → 5 / 0.04 /. s → 0.001 /. x → x1,
  D[-phiChild[x], x] /. consts2 /. ji → 10 / 0.04 /. s → 0.001 /. x → x1,
  D[-phiChild[x], x] /. consts2 /. ji → 20 / 0.04 /. s → 0.001 /. x → x1},
 {x1, 0, .001},
 Frame -> True, Axes → None,
 BaseStyle → {Medium, FontFamily → "Helvetica"},
 FrameLabel -> {"x (m)", "E (V/m)"},
 PlotLabel → "E Field, Child's Law Sheath",
 PlotStyle → {Black},
 Epilog → {Line[{{-10, 0}, {10, 0}}],
   Line[{{0.0010, -300}, {0.0011, -300}}],
   Text["5A", {0.0002, -660 000}],
   Text["10A", {0.0002, -1 100 000}],
   Text["20A", {0.0002, -1 750 000}]]},
 ImageSize → Large]
```

42

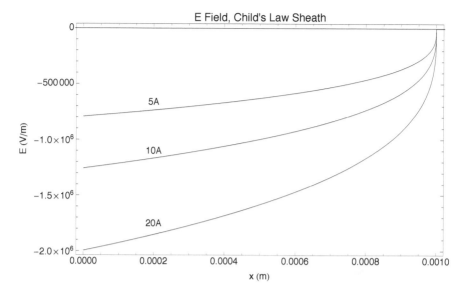

Figure 3- 6. E field in sheath, as predicted by Child's Law.

3.3 TWO FLUID MODEL

Up until now we have assumed that there are no electrons in the sheath. They have all been swept out due to their light mass and the presence of the electric field. In reality, there are electrons present, but not many. Sheridan and Goree (Sheridan and Goree 1989) created a sheath model in which they assumed the electrons are in thermal equilibrium in the sheath.

When the electrons are in thermal equilibrium, they obey the Boltzmann relation (Liberman and Lichtenberg 2005):

$$n_e(x) = n_{e0} exp\left(-\frac{\phi(x)}{k\,T_e}\right) \tag{3-18}$$

where n_e and n_{e0} are the concentration of electrons in the sheath and the bulk, respectively, k is Boltzmann's constant, and T_e is the electron temperature. For the ion concentration, Sheridan and Goree enforced continuity and energy conservation as we did above, but did not neglect the ion velocity at the sheath boundary, v_0.

$$n_i = \frac{J_i}{e\sqrt{v_0^2 - \frac{2e\phi}{m}}} \qquad (3\text{-}19)$$

The Poisson equation now becomes

$$\frac{d^2\phi}{dx^2} = -\frac{e}{\epsilon_0}(n_i - n_e)$$

$$\frac{d^2\phi}{dx^2} = -\frac{en_0}{\epsilon_0}\left[\frac{v_0}{\sqrt{v_0^2 - \frac{2e\phi}{m}}} - exp\left(\frac{e\phi}{kT_e}\right)\right] \qquad (3\text{-}20)$$

There is no analytic solution for this expression. Sheridan and Goree derived an approximate analytic solution. However, we can use Mathematica to numerically integrate equation (3-20), to find the solution.

The commands needed to solve this differential equation are listed in Code 3- 1. Because it is a numerical solution, we need to specify all of the details, like the plasma density and electron temperature, up front. By wrapping the NDSolve function with a With function, we can specify all of these values in a very readable fashion.

Code 3- 1. Solving the Two Fluid model numerically.

```
ans3 = With[{k = 1.381 × 10⁻²³, φw = -1600, Te = 2 eV, e = 1.602 × 10⁻¹⁹,

   ε0 = 8.85 × 10⁻¹², n0 = 10¹⁷, M = mAr, v0 = 1.05 √(2 eV / mAr) , ηw = 800, Ma = 1.05},

   NDSolve[{φ''[u] == - (e n0)/ε0 ((1 - (2 e φ[u])/(M v0²))^(-1/2) - Exp[(e φ[u])/Te]),

   φ[0.00] == φw,

   φ'[0] == (-√(Te n0 / ε0)) (-√2) (Ma² (1 + (2 ηw)/Ma²)^(1/2) - Ma² + Exp[-ηw] - 1)^(1/2)},

   φ, {u, 0, 0.006}, WorkingPrecision → 16]

]
```

Two boundary conditions are specified; the potential at the target surface and the gradient of the potential. The latter is from Sheridan and Goree's paper.

The solution generated by Mathematica is an interpolating function of potential versus distance. Plotting this function, we can see how the potential varies in the sheath.

```
Plot[Evaluate[φ[x] /. ans3], {x, 0, 0.006}, Frame -> True, Axes → None,
  BaseStyle → {Medium, FontFamily → "Helvetica"},
  FrameLabel -> {"Distance from Target (m)", "φ (V)"},
  PlotLabel → "Potential vs Distance",
  PlotStyle → Black, PlotRange → All]
```

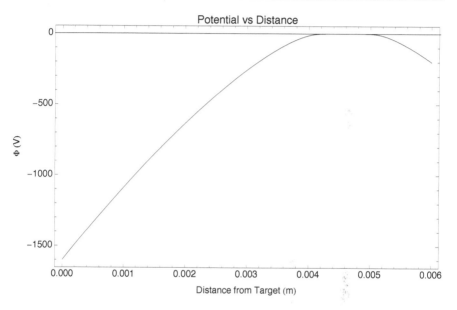

Figure 3- 7. Potential for Two Fluid model. This solution is inaccurate at large distances.

This potential shows the expected behavior over most of the range. It starts at the target potential and monotonically approaches zero at the sheath. However, rather than remaining at zero further into the plasma, the potential turns toward large negative values again. This shows a limitation of the numerical approach. Small errors can build up as we integrate from the wall into the plasma, and these errors can affect the solution. The Mathematica code above for solving Poisson's equation specifies a working precision of 16 digits. If we increase that to 32 digits, we get a much more realistic solution beyond the sheath. In this case the solution time is still short—just a few seconds.

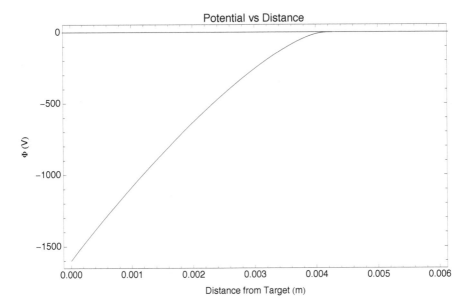

Figure 3- 8. Potential for Two Fluid model with more accurate solution.

It would be nice to compare these three sheath models. To do that, we need to adjust them a bit so they are modeling the same conditions. By setting the current to 9.4A in the Child's Law model, we can match the ion matrix wall voltage. For the Two Fluid model, we specified an ion density of 10^{17} and a velocity at the sheath edge of 1.05 times the Bohm velocity ($v_B = \sqrt{Te/M}$). If we assume the same racetrack area we used for Child's Law, we can calculate the current:

$$I = e\, n_0\, v_0\, A = e\, n_0\, 1.05\, \sqrt{\frac{Te}{M}}\, A$$

$$v_0 = 1.05\, \sqrt{\frac{2eV \times 1.602 \times 10^{-19} J/eV}{6.64 \times 10^{-26} kg}} = 2306 m/s$$

$$I = (1.602 \times 10^{-19} C)(10^{17} \#/m^3)(2306 m/s)(0.04 m^2) = 1.48A$$

The argon ions are entering the sheath at about 2300 m/s. That speed combined with their density give us a current density. The result is a current of about 1.5 A. To scale this up to 9.4 A, we can boost the ion density by the ratio of 9.4 to 1.48. That gives us an ion density of 6.36 x 10^{17}.

Code 3- 2. Two fluid model with current scaled to 9.4A.

$$\text{ans3a} = \text{With}\Bigg[\Bigg\{k = 1.381 \times 10^{-23}, \phi w = -900, Te = 2\,ev, e = 1.602 \times 10^{-19},$$

$$\epsilon 0 = 8.85 \times 10^{-12}, n0 = 6.361 \times 10^{17}, M = mAr, v0 = 1.05\sqrt{\frac{2\,ev}{mAr}}\Bigg\},$$

$$\text{NDSolve}\Bigg[\Bigg\{\phi''[u] == -\frac{e\,n0}{\epsilon 0}\Bigg(\Big(1 - \frac{2\,e\,\phi[u]}{M\,v0^2}\Big)^{-1/2} - \text{Exp}\Big[\frac{e\,\phi[u]}{Te}\Big]\Bigg),$$

$$\phi[0.00] == \phi w,$$

$$\phi'[0] == \sqrt{2}\sqrt{\frac{n0\,Te}{\epsilon 0}}\Bigg(\frac{M\,v0^2}{Te}\Big(1 - \frac{2\,e\,\phi w}{M\,v0^2}\Big)^{1/2} - \frac{M\,v0^2}{Te} + \text{Exp}\Big[\frac{e\,\phi w}{Te}\Big] - 1\Bigg)^{1/2}\Bigg\},$$

$$\phi, \{u, 0, 0.002\}, \text{WorkingPrecision} \to 16\Bigg]$$

$$\Bigg]$$

```
Plot[{phi[x] /. consts,
  phiChild[x] /. consts2 /. ji → 9.4 / 0.04 /. s → 0.001,
  Evaluate[ϕ[x] /. ans3a]
  }, {x, 0, .0015},
Frame -> True, Axes → None,
BaseStyle → {Medium, FontFamily → "Helvetica"},
FrameLabel -> {"x (m)", "ϕ (V)"},
PlotLabel → "Ion Matrix vs Child's vs 2-Fluid",
PlotStyle → {Black, {Black, Dashed}},
PlotRange → {{0, 0.0012}, All},
Epilog → {Line[{{-10, 0}, {10, 0}}],
  Text["Ion Matrix", {0.0002, -400}],
  Text["Child", {0.00050, -300}],
  Text["2-Fluid", {0.00070, -300}]},
ImageSize → Large]
```

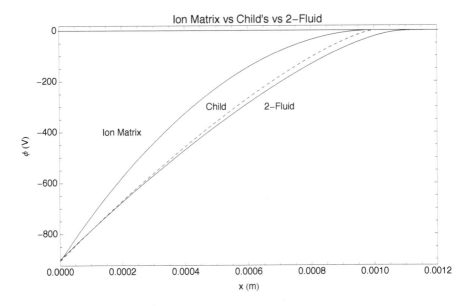

Figure 3- 9. Predicted potential for three different sheath models

The potential predicted by the Ion Matrix model is more curved than that of the Child's Law and Two Fluid models. This is due to the constant ion concentration in the matrix model. Interestingly, the Child's Law and Two Fluid models are very similar in their predicted potential. The largest discrepancy between the two is at the sheath boundary. Since the models differ primarily in the treatment of the electrons, this is telling—namely that there are some electrons present in the sheath near the interface with the plasma, but very few near the wall.

```
Plot[{EField[x1] /. consts,
  D[-phiChild[x], x] /. consts2 /. ji → 9.4 / 0.04 /. s → 0.001 /. x → x1,
  EField3a /. x → x1
  }, {x1, 0, .0015},
  Frame -> True, Axes → None,
  BaseStyle → {Medium, FontFamily → "Helvetica"},
  FrameLabel -> {"x (m)", "E (V/m)"},
  PlotLabel → "Ion Matrix vs Child's vs 2-Fluid",
  PlotStyle → {Black, {Black, Dashed}},
  PlotRange → {{0, 0.0012}, All},
  Epilog → {Line[{{-10, 0}, {10, 0}}],
    Text["Ion Matrix", {0.0003, -1.6 10^6}],
    Text["Child", {0.0002, -1.22 10^6}],
    Text["2 Fluid", {0.0002, -1.02 10^6}]},
  ImageSize → Large]
```

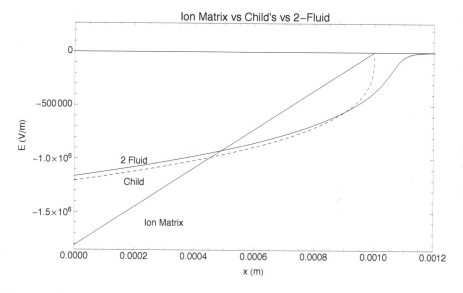

Figure 3- 10. Predicted electric field for three different sheath models

The electric field reveals more differences between these models. As noted earlier, the Ion Matrix model has a linear E field. The Child's Law E field is flatter through much of the sheath, but drops off rapidly near the sheath-plasma interface. Meanwhile, the Two Fluid model follows Childs Law through much of the sheath, but the presence of electrons near the sheath boundary smoothes the transition to zero E field.

3.4 MAGNETIZED SHEATH

Up until now we have ignored the effect of the magnetic field on the sheath. This is a large omission, but a necessary one in order to simplify the problem. Because the Lorentz force is perpendicular to the electron velocity, it is very difficult to create a one-dimensional sheath model that includes magnetic fields. There have been a few efforts, for instance (Lister 1996), (Bradley, Thompson and Gonzalvo 2001), and (Cramer 1997), but the accuracy of these models is uncertain. The most accurate method for modeling the magnetized sheath is through Particle-in-Cell Monte Carlo (PIC-MC) simulations (Bultinck, et al. 2010). These are outside the scope of this book. Instead we will outline the method pursued by Gu and Lieberman (Gu and Lieberman 1988).

Rather than starting with Poisson's equation, Gu and Lieberman collected experimental data on a cylindrically symmetric magnetron system. Using optical emission spectroscopy, they measured the brightness of the plasma as a function of distance from the target. The distance to the point of peak brightness varied with B field strength, current, and pressure. They reasoned that this distance could be correlated with one of two length scales: the gyroradius of the energetic electrons or the width of the Child's Law sheath. If it was due to the gyroradius, the distance would scale as

$$r_L = \frac{mv_\perp}{eB} \tag{3-21}$$

The electron's velocity would be that acquired falling through the sheath potential and thus would be $v = (2eV/m)^{1/2}$. The radius then becomes

$$r_L = \frac{2m^{1/2}}{e} \frac{V^{1/2}}{B} \tag{3-22}$$

If instead the peak distance varied with Child's Law then the dependence would follow equation (3-17):

$$d = \frac{2}{3}\epsilon_0^{1/2} \left(\frac{2e}{m}\right)^{1/4} \frac{V^{3/4}}{J^{1/2}}$$

As noted in (Liberman and Lichtenberg 2005) page 562, the width of the racetrack for a disk-shaped target depends on the Larmor radius and the curvature of the field lines, a:

$$w \approx \sqrt{2ar_L}$$

If the racetrack is at radius r_0, we can then calculate the area over which sputtering occurs and use that to convert the current density J into a current I. Our expression for the sheath thickness then becomes

$$d = \frac{4\sqrt{2}}{3}(\pi r_0 \epsilon_0)^{1/2} \left(\frac{e}{m}\right)^{1/4} \left(\frac{2m}{e}\right)^{1/8} \frac{a^{1/4}V^{7/8}}{I^{1/2}B^{1/4}} \qquad (3\text{-}23)$$

Gu and Lieberman found that their experimental data more closely followed the voltage and B field dependence of equation (3-23). From this they concluded that the peak in brightness correlated with the sheath thickness and the thickness goes as $s{\sim}V^{7/8}B^{-1/4}$. Thus, the sheath thickness increases with increasing voltage and decreases with increasing B field.

While this analysis builds our intuitive understanding of the sheath it doesn't provide any information on the electric field in the sheath. In the next chapter, when we need this field to move electrons, we will use the linear field of the ion matrix model.

4. ELECTRON MOTION

With the groundwork laid in the last two chapters, we can now simulate the motion of electrons in the presence of electric and magnetic fields. The right combination of E and B fields will trap the electrons near the target surface, creating high electron densities and consequently high sputter rates.

4.1 MOTION WITH TWO MAGNETS

The electron motion is governed by the Lorentz force law:

$$F = qE + qv{\times}B \qquad (4\text{-}1)$$

where F is the force on the electron and q is the electron charge. We know from Newton's second law that force equals mass times acceleration. Using this, along with the fact that the acceleration is the second derivative of position, we can write equation (4- 1) as

$$m\ddot{x} = qE + qv{\times}B \qquad (4\text{-}2)$$

where $\ddot{x}$ is the second derivative of position with time. This is a vector equation, and for us it will be more convenient to write it as three equations, one for each direction. If we don't remember the definition of a cross product, Mathematica can remind us:

```
{vx, vy, vz} × {Bx, By, Bz}
```

```
{Bz vy - By vz, -Bz vx + Bx vz, By vx - Bx vy}
```

The electron equations of motion are then

$$m\ddot{x} = qE_x + q\left(\dot{y}B_z - \dot{z}B_y\right) \qquad (4\text{-} 3)$$

$$m\ddot{y} = qE_y + q\left(\dot{z}B_x - \dot{x}B_z\right) \qquad (4\text{-} 4)$$

$$m\ddot{z} = qE_z + q\left(\dot{x}B_y - \dot{y}B_x\right) \qquad (4\text{-} 5)$$

where the first derivative of position is used to represent the velocity components. By solving these three coupled differential equations, we can determine the position of a single electron over time. The Mathematica function NDSolve is designed just for this type of problem. Let's try it out with a simplified B field, rather than a full magnet array. Consider two long parallel magnets, as shown in Figure 4- 1.

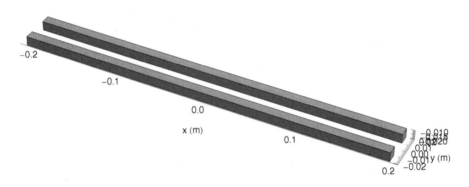

Figure 4- 1. Simplified magnet array.

We will set each magnet to be 1 cm x 1cm x 40 cm in size with a magnetization of Br = 1.4 T. Using our methodology from Chapter 2, we can write this as

Code 4- 1. Definition of 2 magnet array.
```
magpack = {
    {{-20 cm, -2 cm, -2 cm}, {40 cm, 1 cm, 1 cm}, Br}, (*left mag*)
    {{-20 cm, 1 cm, -2 cm}, {40 cm, 1 cm, 1 cm}, -Br}(*right mag*)
};
```

With the B field functions from Chapter 2 (Code 2- 4), we can plot the field. The upper surface of the target is imagined to be at z=0, indicated by the horizontal line in Figure 4- 2. The voltage applied there will create our E field.

Code 4- 2. StreamPlot is used to plot the B field vectors.

```
drawMag2d[magpack_, magNum_] := Rectangle[magpack[[magNum, 1, 2 ;; 3]],
    magpack[[magNum, 1, 2 ;; 3]] + magpack[[magNum, 2, 2 ;; 3]]]
```

```
p0 = StreamPlot[{By[{0, y, z}, magpack],
    Bz[{0, y, z}, magpack]}, {y, -0.03, 0.03},
    {z, -0.02, 0.025}, BaseStyle → {Medium, FontFamily → "Helvetica"},
    AspectRatio → 0.045 / 0.06,
    StreamStyle → Black,
    Prolog → {Gray, drawMag2d[magpack, 1], drawMag2d[magpack, 2]},
    Epilog → {Thick, Black, Line[{{-0.03, 0}, {0.03, 0}}]},
    FrameLabel → {"y (m)", "z (m)"}, ImageSize → Large]
```

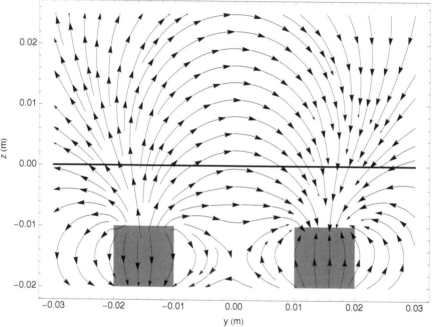

Figure 4- 2. Field lines from two magnets.

From Equation (3-5), the field varies linearly with distance from the target. In Mathematica we can write:

Code 4- 3. Definition of linear E field.

```
Clear[Ex, Ey, Ez]
```

```
Ex[x_, y_, z_, Vd_, d_] := 0
```

```
Ey[x_, y_, z_, Vd_, d_] := 0
```

$$Ez[x\_, y\_, z\_, Vd\_, d\_] := \text{Piecewise}\left[\left\{\left\{\frac{2\,Vd}{d}\,\frac{d-z}{d}, z < d\right\}\right\}, 0\right]$$

The E field is zero in the x and y directions. In the Mathematica code for the z direction, we specify the field using the Piecewise function. This allows us to distinguish the sheath and nonsheath regions. Alternatively, an If statement can be used, but this can cause problems when using NDSolve.

We will assume the target is set to -300 V and the sheath is 1mm thick. Near the target, the electric field reaches -600,000 V/m, and then linearly drops to zero at the edge of the sheath.

Code 4- 4. Plotting the electric field.

```
Plot[Ez[0, 0, z, -300, 0.001], {z, 0.0, 0.002},
 Frame -> True,
 Axes → None,
 BaseStyle → {14, FontFamily → "Helvetica"}, FrameLabel -> {"s (m)", "E (V/m)"},
 PlotStyle → Black,
 PlotLabel → "E Field",
 Epilog → {Black, Dashed, Line[{{0.001, -10^6}, {0.001, 1}}],

  Text["Sheath\nEdge", {0.00114, -5 10^5}]},
 ImageSize → Large, PlotRange → All]
```

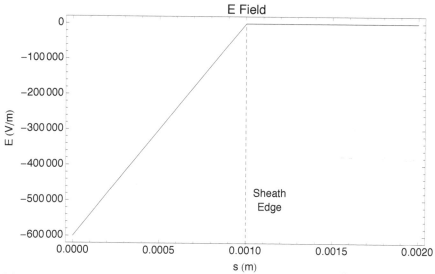

Figure 4- 3. Linear electric field in sheath.

We can now solve our differential equations by calling the NDSolve function:

Code 4- 5. Solving the electron equations of motion.

```
(ans1 = NDSolve[{
    m D[x[t], t, t] ==
        -q * (Ex[x[t], y[t], z[t], -300, .001] + D[y[t], t] * Bz[{x[t], y[t], z[t]},
            magpack] - D[z[t], t] * By[{x[t], y[t], z[t]}, magpack]),
    m D[y[t], t, t] == -q * (Ey[x[t], y[t], z[t], -300, .001] - D[x[t], t] * Bz[{x[t],
            y[t], z[t]}, magpack] + D[z[t], t] * Bx[{x[t], y[t], z[t]}, magpack]),
    m D[z[t], t, t] == -q * (Ez[x[t], y[t], z[t], -300, .001] + D[x[t], t] * By[{x[t],
            y[t], z[t]}, magpack] - D[y[t], t] * Bx[{x[t], y[t], z[t]}, magpack]),
    x[0] == 0, y[0] == -0.007, z[0] == 0.000,
    x'[0] == v0[[1]], y'[0] == v0[[2]], z'[0] == v0[[3]]},
    {x, y, z},
    {t, 0.0, t1},
    MaxStepSize → 10⁻¹¹, MaxSteps → 500 000,
    AccuracyGoal → 13, PrecisionGoal → 13,
    WorkingPrecision → 20, EvaluationMonitor :→ (currTime = t;)]) // Timing
```

The first portion of the NDSolve expression lists our three equations of motion, for the x, y, and z directions. The next section has the initial conditions for the electron. In our case these are the starting position at the target surface (0, -0.007, 0) and the starting velocity (0, 0, 0). Next we list the variables we are solving for (x, y, z) and the range of times we are solving over. The last section has various solver settings. These are used to guide the solver to find a solution. In our case, boosting the accuracy and precision goals leads to a more accurate solution.

It takes a few minutes to solve this expression. Once it is done, Mathematica returns an interpolating function for each variable. We can plot these using a parametric plot to see the trajectory the electron takes over the target.

Code 4- 6. Electron trajectory plotted using ParametricPlot
```
p1 = ParametricPlot3D[Evaluate@({x[t], y[t], z[t]} /. ans1[[1]]),
    {t, 0, t1}, PlotRange → {All, {-0.015, 0.015}, All},
    ImageSize → {400, 400}, BoxRatios → {10, 2, 0.2},
    PlotStyle → Black, BaseStyle → {14, FontFamily → "Helvetica"},
    AxesLabel → {Style["x", Italic], Style["y", Italic], Style["z", Italic]}]
```

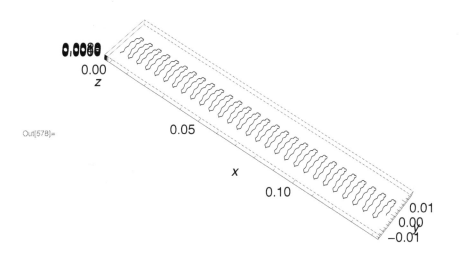

Out[578]=

Figure 4- 4. Electron trajectory.

The electron follows a zigzag path, drifting in the x direction. The side view of this trajectory can be seen in Figure 4- 5, superimposed over the B field.

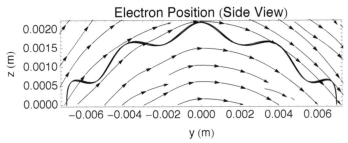

Figure 4- 5. Side view of electron trajectory.

The electron leaves the target surface almost vertically at y = -0.007 m and then begins circling around a B field line until it reaches the target again. The E field then repels it and it retraces its path to where it started. In this way, the electron is trapped near the target surface. In addition to this back-and-forth motion in the y-z plane, there is the ExB drift in x direction that takes the electron down the magnetron.

To ensure our solution is accurate, we should confirm that energy is conserved. At any given point the total energy of the electron consists of its kinetic energy and its potential energy. The latter comes about from its position in the electric field. Only when in the sheath will the electron have any potential energy.

The kinetic energy is given by

$$KE = \frac{1}{2}m\left(v_x^2 + v_y^2 + v_z^2\right) \qquad (4\text{-}6)$$

The potential energy can be found by integrating the force on the electron as it moves from its current position to the sheath boundary, z = s.

$$PE = \int_z^s q\,E_z\,dz \qquad (4\text{-}7)$$

We can have Mathematica do this integration, using the Integrate function:

```
pe = Integrate[-q0 Ez[0, 0, zz, Vd, d], {zz, z, d}, Assumptions → {z, d} ∈ Reals]
```

$$\begin{bmatrix} \frac{-d^2\,q0\,Vd + 2\,d\,q0\,Vd\,z - q0\,Vd\,z^2}{d^2} & d - z > 0 \\ 0 & True \end{bmatrix}$$

The results can be simplified with the Collect function:

$$pe = \text{Collect}\left[\frac{-d^2\ q0\ Vd + 2\ d\ q0\ Vd\ z - q0\ Vd\ z^2}{d^2}, \{q0,\ Vd\}\right]$$

$$\frac{q0\ Vd\ \left(-d^2 + 2\ d\ z - z^2\right)}{d^2}$$

The expression can be made simpler still by writing it in terms of E instead of V. This can be done by taking our expression for Ez, solving it for Vd, and substituting that expression into the potential energy expression above. An additional simplification using the Apart and Simplify functions gets us to the final form:

$$eq1 = \text{Solve}\left[Ez == \frac{2\ Vd}{d}\ \frac{d - z}{d}, Vd\right]$$

$$\left\{\left\{Vd \rightarrow \frac{d^2\ Ez}{2\ (d - z)}\right\}\right\}$$

```
eq2 = pe /. eq1
```

$$\left\{\frac{Ez\ q0\ \left(-d^2 + 2\ d\ z - z^2\right)}{2\ (d - z)}\right\}$$

```
Simplify[Apart[eq2]]
```

$$\left\{\frac{1}{2}\ Ez\ q0\ (-d + z)\right\}$$

Thus the potential energy can be written as

$$PE = \frac{1}{2}q_0 E_z(z - s) \tag{4-8}$$

This, combined with the kinetic energy, can be used to plot the total energy, as shown in Code 4-7.

Code 4-7. Plotting the total electron energy.

```
Plot[

  Evaluate@ ( 1/2 m (x'[tt]^2 + y'[tt]^2 + z'[tt]^2) + (pe /. {z → z[tt], d → 0.001, Vd → -300,

       q0 → q}) /. ans1[[1]] ) / q, {tt, 0, 10 10^-8},

  Frame -> True, Axes → None,
  BaseStyle → {Medium, FontFamily → "Helvetica"},
  FrameLabel -> {"t (s)", "Total Energy (eV)"},
  PlotLabel → "", PlotRange → {All, {0, 500}},
  PlotPoints → 1000, PlotStyle → Black, ImageSize → Large]
```

In Figure 4- 6, we can see that the total energy is conserved. This confirms that NDSolve is giving a reasonable result. We can also plot the potential and kinetic energy individually and see how the energy cycles back and forth between the two (Figure 4- 7 and Figure 4- 8). Each time the electron returns to the target surface, all of its energy is converted into potential energy, and its speed drops to zero.

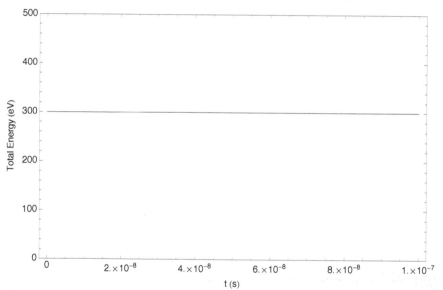

Figure 4- 6. Total energy of the electron is conserved

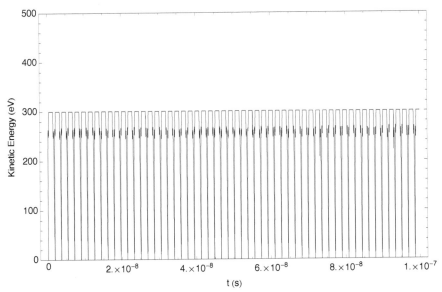

Figure 4- 7. Kinetic energy of electron.

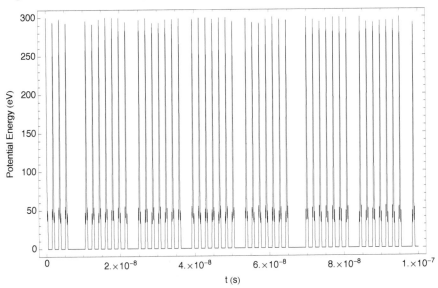

Figure 4- 8. Potential energy of electron.

After an electron is emitted from the target surface, it quickly accelerates to high speeds. This can be seen in Figure 4- 9. With -300 V applied to the target, the electron reaches speeds of over 8 x 10⁶ m/s in less than one nanosecond. While this is quite fast, it is still just a few percent of the speed of light. That is what allows us to ignore relativistic effects and use classical physics in solving for the electron trajectory.

Code 4- 8. A ParametricPlot is used to plot electron position vs velocity.

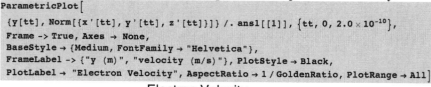

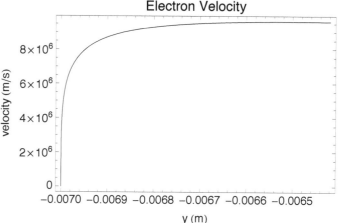

Figure 4- 9. Electron velocity after leaving target surface.

One can make some very interesting figures by plotting the phase space coordinates. For instance in Figure 4- 10, the y component of velocity is plotted against the y position. In Figure 4- 11, the same is done for the x and z components.

```
p2 = ParametricPlot[Evaluate@({y[tt], y'[tt]} /. ans1[[1]]), {tt, 0, 10⁻⁸},
    Frame -> True, Axes → None,
    BaseStyle → {Medium, FontFamily → "Helvetica"},
    FrameLabel -> {"y (m)", "v_y (m/s)"}, PlotStyle → Black,
    PlotLabel → "Phase Space - v_y vs y", AspectRatio → 6 / 20]
```

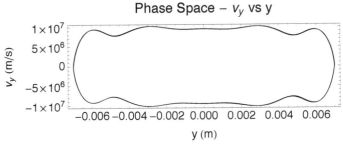

Figure 4- 10. Y position versus y velocity.

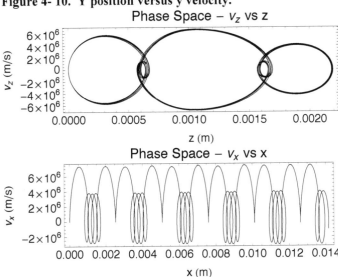

Figure 4- 11. Z and x positions versus z and x components of velocity.

4.2 DRIFT VELOCITY

A key aspect of magnetron sputtering is the fact that electrons follow a closed loop as they move over the target surface. This is due to their drift velocity. The first thing to point out about the drift velocity is that different electrons have different drift rates. Those in the center of the racetrack go much faster. Those further away from the center line of the racetrack drift more slowly. We can see that by launching some additional electrons.

In Figure 4- 12, we launch from three different y positions, -0.007 m, -0.015 m, and -0.020 m, rerunning NDSolve for 0.1 μs for each case.

Code 4- 9. ParametricPlot3D used on a list of solutions.

```
p1 = ParametricPlot3D[{Evaluate@({x[t], y[t], z[t]} /. ans1[[1]]),
    Evaluate@({x[t], y[t], z[t]} /. ans2[[1]]),
    Evaluate@({x[t], y[t], z[t]} /. ans3[[1]])}, {t, 0, 1×10⁻⁷},
  PlotRange → {{-.0, 0.2}, {-0.03, 0.03}, All},
  ImageSize → {400, 400}, BoxRatios → {10, 2, 0.2}, PlotStyle → Black,
  AxesLabel → {Style["x", Italic], Style["y", Italic], Style["z", Italic]}]
```

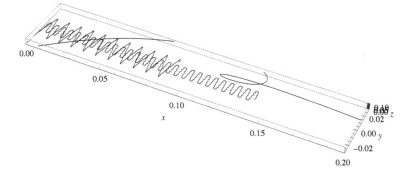

Figure 4- 12. Launching electrons from three starting locations.

The inner trajectory is our original launch point. By moving the launch point out to 0.015 m, the trajectory is wider, and the drift velocity is much lower. Going further out to 0.020 m, the electron is only weakly bound by the magnetic field and doesn't seem to orbit the B field lines. This can be seen more clearly in a side view (Figure 4- 13).

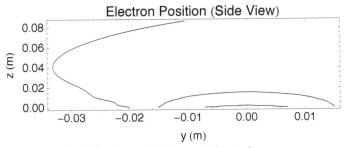

Figure 4- 13. Side view of electron trajectories.

This electron drift down the racetrack is generally referred to as ExB drift. However, other mechanisms also contribute to the drift. In addition to ExB drift, there is gradient drift and curvature drift (Chen 1984). All three cause the electron to move in the x direction. We can estimate the magnitude of these mechanisms to get a feel for which dominates. The average drift velocity of the electron launched from y = -0.07 m is

$$v_{drift} = \frac{1}{t} \int v_x \, dt$$

```
vDrift =  NIntegrate[x'[tt] /. ans1[[1]], {tt, 0, 10⁻⁷}]
          ────────────────────────────────────────────────  // Quiet
                              10⁻⁷
```

```
1.40157 × 10⁶
```

On average, the electron is moving at 1.4 x 10^6 m/s in the x direction. The ExB drift is given by

$$v_{ExB} = \frac{E \times B}{B^2}$$

We can plot this for our electron and see how it varies with position.

Code 4- 10. Position versus ExB drift velocity.

```
p20 =
  ParametricPlot[Evaluate@({x[t], ({0, 0, Ez[x[t], y[t], z[t], -300, 0.001]} × {0,
        By[{x[t], y[t], z[t]}, magpack], 0} / By[{x[t], y[t], z[t]}, magpack]²)[[
        1]]} /. ans1[[1]]), {t, 0, 1.08 × 10⁻⁸},
    Frame -> True, Axes → None,
    BaseStyle → {Medium, FontFamily → "Helvetica"},
    FrameLabel -> {"x (m)", "vₓ (m/s)"}, PlotStyle → Black,
    PlotLabel → "E x B Drift Velocity", AspectRatio → 1 / GoldenRatio]
```

E x B Drift Velocity

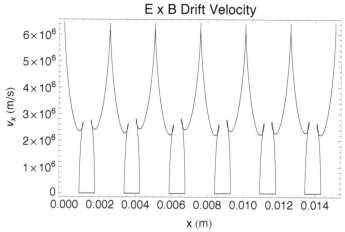

Figure 4- 14. ExB drift velocity.

As expected, the ExB drift velocity can be quite high when the electron is in the sheath, reaching more than 6×10^6 m/s. But outside of the sheath it goes to zero. The average ExB drift velocity is calculated as follows:

Code 4- 11. Calculating average ExB drift.

```
vExBAve =
 NIntegrate[({0, 0, Ez[x[t], y[t], z[t], -300, 0.001]} ×{0, By[{x[t], y[t], z[t]},
     magpack], 0}/By[{x[t], y[t], z[t]}, magpack]² /.
  ansl[[1]])[[1]], {t, 0, 1.08 × 10⁻⁸}] / (1.08 × 10⁻⁸) // Quiet
```

```
1.36764 × 10⁶
```

This is quite similar to the average drift velocity seen by our electron, which suggests ExB drift is the dominant mechanism. This is consistent with experimental measurements of Bradley, et al. (Bradley, Thompson and Gonzalvo 2001)

Even though the ExB drift drops to zero outside the sheath, our plot of trajectory (Figure 4- 11) shows an apparent drift in the x direction all of the time. This suggests that the other drift mechanisms do play some role as well.

The other two sources of electron drift in a B field are gradient drift and curvature drift. The grad B drift is given by (Chen 1984):

$$v_{gradB} = \frac{1}{2} v_\perp r_L \frac{B \times \nabla B}{B^2}$$

where $v_\perp$ is the electron velocity perpendicular to the field and r_L is the Larmor radius given by

$$r_L = \frac{mv_\perp}{qB}$$

The curvature drift is given by

$$v_{curve} = \frac{mv_\parallel}{qB} \frac{r_c \times B}{r_c^2 B}$$

where $v_\parallel$ is the electron velocity parallel to the B field and r_c is the radius of curvature of the B field.

The Grad B drift can be calculated with this function:

Code 4- 12. A function for the Grad B drift velocity.

```
vGradBFunc[mags_, t1_] := Module[{B1, gradB, vPerp},
  B1 = B[{x[t], y[t], z[t]}, mags] /. t → t1 /. ans1[[1]];
  (*Print[B1];*)
  gradB =
    Grad[By[{x[t], y[t], z[t]}, mags], {x[t], y[t], z[t]}] /. t → t1 /. ans1[[1]];
  (*Print[gradB];*)
  vPerp = (v - v.Normalize[B1]) /. t → t1 /. ans1[[1]];
  1          m vPerp    B1 × gradB
  ─ vPerp ─────────────── ───────────
  2          -q Norm[B1]   Norm[B1]²

]
```

Plotting this drift (Figure 4- 15) we see that is it quite small compared to ExB drift. On average the curvature drift is only

```
vDriftAve = NIntegrate[vGradBFunc[magpack, t][[1]] /. ans1〚1〛,
  {t, 0, 1.08 × 10⁻⁸}, AccuracyGoal → 4] / (1.08 × 10⁻⁸)
```
202772

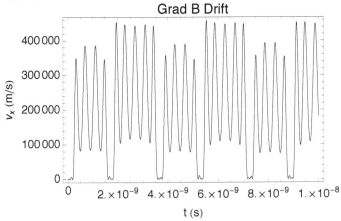

Figure 4- 15. Velocity due to gradient drift.

For the curvature drift, we need to estimate the curvature of the B field. From Figure 4- 2 we can see that for an electron starting at y = 0.007 m, the curvature is about 1.5 cm. A more precise calculation shows the average curvature of the B field along this trajectory to be 1.38 cm. The radius vector can then be written as

$$rC = \{0, \; r_{ave} \frac{-B_z}{\sqrt{B_y^2 + B_z^2}}, \; r_{ave} \frac{B_y}{\sqrt{B_y^2 + B_z^2}}\}$$

The curvature drift can then be calculated with this expression:

$$\text{vCurve} = \frac{m\, \text{vPara}^2}{-q\,\text{Norm}[B[\{x[t], y[t], z[t]\}, \text{magpack}]]^2} \; \frac{rC \times B[\{x[t], y[t], z[t]\}, \text{magpack}]}{\text{Norm}[rC]^2};$$

A plot of the curvature drift is shown in Figure 4- 16.

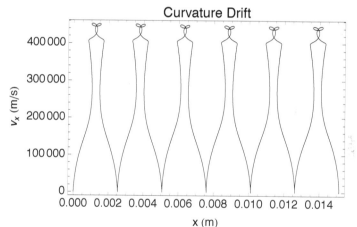

Figure 4- 16. Velocity due to curvature drift.

The peak curvature drift is about half the size of the ExB drift. More importantly, this peak drift occurs exactly when the ExB drift is zero. This can be seen by overlaying the two plots:

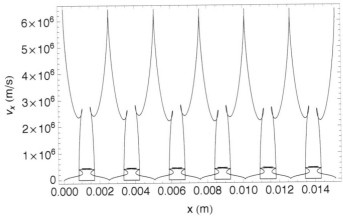

Figure 4- 17. Curvature drift overlaid on ExB drift.

The average curvature drift is

```
vCurveAve =
  NIntegrate[Evaluate@ (vCurve[[1]] /. ans1), {t, 0, 1.08 × 10⁻⁸}] / (1.08 × 10⁻⁸)

{156319.}
```

From the average drift values we can say that for this case, the electron drift is due primarily to ExB, with small contributions from Grad B drift and curvature drift. We made a few simplifying assumptions and as a result the three mechanisms don't add up to the average drift velocity calculated at the start of this section. Still, these calculations help us understand the nature of the drift of electrons around the racetrack.

4.3 VARYING MAGNET STRENGTH

In practice it is difficult to maintain a constant magnetic field strength around the entire racetrack. In particular, the turnaround region typically has a weaker field. What is the effect of magnetic field gradients on the electron trajectory? Buyle et al. found that the height, width and velocity of the electron trajectory all change as the electron transitioned from a region of weak field to high field (Buyle, et al. 2004). We can do a similar calculation by modifying our magpack list slightly:

```
magpack = {
    {{-20 cm, -2 cm, -2 cm}, {20 cm, 1 cm, 1 cm}, Br / 2}, (*left mag weak*)
    {{0 cm, -2 cm, -2 cm}, {20 cm, 1 cm, 1 cm}, Br}, (*left mag strong*)
    {{-20 cm, 1 cm, -2 cm}, {20 cm, 1 cm, 1 cm}, -Br / 2}, (*right mag weak*)
    {{0 cm, 1 cm, -2 cm}, {20 cm, 1 cm, 1 cm}, -Br}(*right mag strong*)
};
```

We have taken each long magnet and divided it into two parts, with a weak end and a strong end. Figure 4- 18 shows the B field in the target plane half way between the magnets.

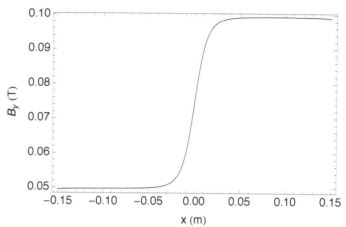

Figure 4- 18. B field strength along length of magnets.

In order to find the electron trajectory, we run the NDSolve function like before. It takes a little longer to run because our magpack has twice the magnets in it.

Code 4- 13. Solving electron equations of motion.

```
(ans1 = NDSolve[{
    m D[x[t], t, t] ==
      -q * (Ex[x[t], y[t], z[t], -300, .001] + D[y[t], t] * Bz[{x[t], y[t], z[t]},
        magpack] - D[z[t], t] * By[{x[t], y[t], z[t]}, magpack]),
    m D[y[t], t, t] == -q * (Ey[x[t], y[t], z[t], -300, .001] - D[x[t], t] * Bz[{x[t],
        y[t], z[t]}, magpack] + D[z[t], t] * Bx[{x[t], y[t], z[t]}, magpack]),
    m D[z[t], t, t] == -q * (Ez[x[t], y[t], z[t], -300, .001] + D[x[t], t] * By[{x[t],
        y[t], z[t]}, magpack] - D[y[t], t] * Bx[{x[t], y[t], z[t]}, magpack]),
    x[0] == -0.1, y[0] == -0.007, z[0] == 0.000,
    x'[0] == v0[[1]], y'[0] == v0[[2]], z'[0] == v0[[3]]},
    {x, y, z},
    {t, 0.0, 1 × 10^-7},
    MaxStepSize → 10^-11, MaxSteps → 500 000,
    AccuracyGoal → 13, PrecisionGoal → 13,
    WorkingPrecision → 20, EvaluationMonitor :→ (currTime = t;)]) // Timing
```

The trajectory is shown in Figure 4- 19. In the weak field region, the electron has a higher drift velocity, so the pathline is more spread out. As the electron moves into the stronger field, its drift velocity slows and the pattern becomes tighter. This means that the electron spends less time in the weak field region. This leads to less time ionizing argon atoms and thus a lower sputter rate in the weak region.

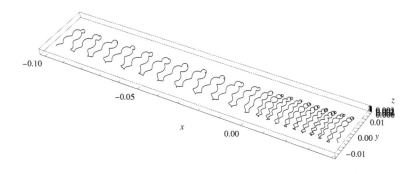

Figure 4- 19. Effect of B Field gradient on electron trajectory.

A side view of the trajectories can be seen in Figure 4- 20. The particle initially follows an arc close to the surface (solid line). Much of the time the electron is in the 1 mm sheath. Once it transitions into the stronger field, the electron moves up to a higher B field line (dashed line). This both broadens its arc and gets it out of the sheath for much of the time. Both of these have implications for sputtering. The broader arc should result in a wider erosion groove in the target, boosting target utilization. The higher trajectory means that more ions are formed above the sheath. As they are attracted to the target, the ions fall through the full sheath potential (300 V in this case), transmitting maximum energy to the target. When ionization occurs inside the sheath, the ions fall through only a portion of the sheath region and thus don't pick up the full 300 eV of energy.

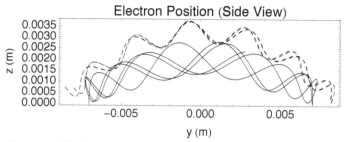

Figure 4- 20. The electron move up to a higher trajectory as the B field strengthens.

We can easily try the reverse case, where the electron starts in a strong field and transitions to a weak field. The magpack then becomes

```
magpack = {
    {{-20 cm, -2 cm, -2 cm}, {20 cm, 1 cm, 1 cm}, Br}, (*left mag strong*)
    {{0 cm, -2 cm, -2 cm}, {20 cm, 1 cm, 1 cm}, Br / 2}, (*left mag weak*)
    {{-20 cm, 1 cm, -2 cm}, {20 cm, 1 cm, 1 cm}, -Br}, (*right mag*)
    {{0 cm, 1 cm, -2 cm}, {20 cm, 1 cm, 1 cm}, -Br / 2} (*right mag*)
};
```

The trajectory plots are shown in Figure 4- 21. In this case, the electron starts in a trajectory with a slow drift velocity and then transitions into a faster one. The arc also goes in the opposite direction. It starts high, and as the field weakens, the electron moves down to a lower B field line.

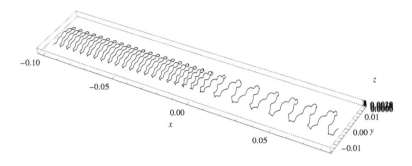

Figure 4- 21. As the B field weakens, the drift velocity increases.

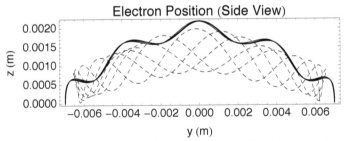

Figure 4- 22. The weakening B field forces the electrons closer to the target surface.

Buyle (Buyle, et al. 2004) used similar results to explain the cross corner effect—that is the observation of higher erosion rates just after the turnaround, at both ends of the target. They noted that as the electron comes out of the turnaround into a stronger field, it both slows down and drops to a lower orbit. Both effects increase the electron density there, leading to more ionization and higher erosion rates.

4.4 FULL MAGNETRON

Using the magnetron from Section 2.2, let's model the electron as it makes its way around the full magnetron. We define out magnet array like before:

```
magpack = {
    {{-17.5 cm, -1 cm, -2 cm}, {35 cm, 2 cm, 1 cm}, Br}, (*center mags*)
    {{-20 cm, -2.5 cm, -2 cm}, {1 cm, 5 cm, 1 cm}, -Br}, (*bottom row*)
    {{-20 cm, 2.5 cm, -2 cm}, {40 cm, 1 cm, 1 cm}, -Br}, (*right side*)
    {{19 cm, -2.5 cm, -2 cm}, {1 cm, 5 cm, 1 cm}, -Br}, (*top row*)
    {{-20 cm, -3.5 cm, -2 cm}, {40 cm, 1 cm, 1 cm}, -Br}}(*left side*)
```

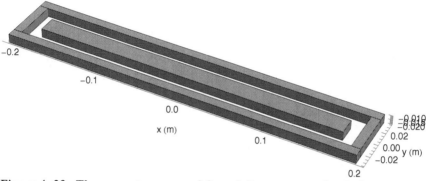

Figure 4- 23. The magnet array used for a full magnetron simulation.

The magnetic field lines in the middle portion of the array can be easily plotted:

Code 4- 14. Using StreamPlot for B field vectors
```
vp2 =
  StreamPlot[{By[{0, y, z}, magpack], Bz[{0, y, z}, magpack]}, {y, -3.5 cm, 3.5 cm},
    {z, -2 cm, 3 cm}, AspectRatio → 5 / 7,
    BaseStyle → {Medium, FontFamily → "Helvetica"},
    StreamStyle → Black,
    PlotLabel → "B Field Vectors", FrameLabel → {"y (m)", "z (m)"},
    Epilog → {Prepend [Map[Rectangle[#[[1, 2 ;; 3]], #[[1, 2 ;; 3]] + #[[2, 2 ;; 3]]] &,
      {magpack[[1]], magpack[[3]], magpack[[5]]}], Gray],
      Thick, Black, Line[{{-0.03, 0}, {0.03, 0}}]},
    ImageSize → Large]
```

75

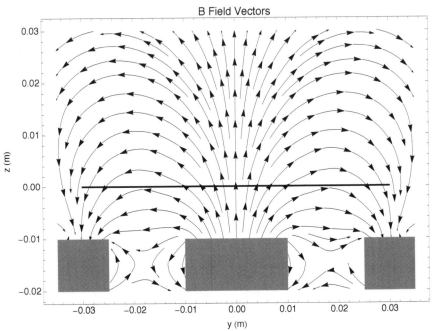

Figure 4- 24. Magnetic field vectors at x = 0. The horizontal line represents the target surface.

We can launch a particle from anywhere along the racetrack and then solve for its position in the usual way. Let's see how the motion evolves as the electron moves through the turnaround region.

Code 4- 15. NDSolve is used to find the electron trajectory at the end of the magnet array.

```
(ans1 = NDSolve[{
    m D[x[t], t, t] ==
        -q * (Ex[x[t], y[t], z[t], -300, .001] + D[y[t], t] * Bz[{x[t], y[t], z[t]},
            magpack] - D[z[t], t] * By[{x[t], y[t], z[t]}, magpack]),
    m D[y[t], t, t] == -q * (Ey[x[t], y[t], z[t], -300, .001] - D[x[t], t] * Bz[{x[t],
            y[t], z[t]}, magpack] + D[z[t], t] * Bx[{x[t], y[t], z[t]}, magpack]),
    m D[z[t], t, t] == -q * (Ez[x[t], y[t], z[t], -300, .001] + D[x[t], t] * By[{x[t],
            y[t], z[t]}, magpack] - D[y[t], t] * Bx[{x[t], y[t], z[t]}, magpack]),
    x[0] == -0.1, y[0] == -0.007, z[0] == 0.000,
    x'[0] == v0[[1]], y'[0] == v0[[2]], z'[0] == v0[[3]]},
    {x, y, z},
    {t, 0.0, t1},
    MaxStepSize → 10^-11, MaxSteps → 500000,
    AccuracyGoal → 13, PrecisionGoal → 13,
    WorkingPrecision → 20, EvaluationMonitor :→ (currTime = t;)}]) // Timing
```

As shown in Figure 4- 25, the electron trajectory has three distinct motions. On a fine scale the electron is circling around the local B field line. On a bigger scale, it is following an arc-shaped path from the target surface upward and then back down to the target. Lastly, it is following the ExB drift direction, which takes it around the racetrack.

This last motion can be seen in Figure 4- 26 where the trajectory is superimposed on both the racetrack (dashed line is Bz = 0) and the contours of the parallel B field. There are several interesting things to note in this figure. First, the electron position in the B field shifts dramatically as it reaches the end of the magnet array. At the start of its trajectory, the electron is primarily in the strongest part of the B field (indicated by lighter shading). In the turnaround region, not only is the field weaker overall, the electron drifts into an even weaker portion of the field. As we saw earlier, this weaker B field results in a faster ExB drift, leading to less time spent in the turnaround region and less ionization and sputtering. So just from this plot we would expect the target erosion to be less at the turnaround than in the straight section of the magnet array.

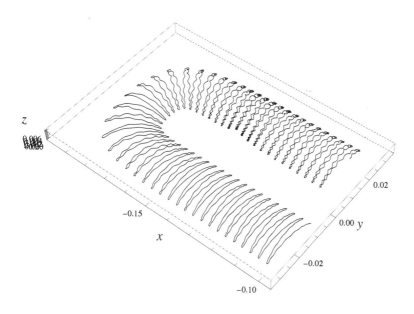

Figure 4- 25. Trajectory of an electron starting at (-0.10, -0.007).

77

Another interesting aspect of the trajectory is the relative positions of the electron trajectory and the dashed Bz = 0 line. The general rule of thumb is that the racetrack will be centered around this line. This is true when the field lines follow nice, symmetric arcs, as in Figure 4- 2. However, the field lines for this magnet array are not symmetric, as shown in Figure 4- 24. At the target surface, the Bz = 0 point is closer to the outer magnets, rather than centered between the inner and outer magnets. A slice through the turnaround region (not plotted) would show the opposite trend. From this we can conclude that the racetrack is only approximately located by the Bz = 0 line.

The electron motion around the field lines also changes as the electron moves through the turnaround. The spirals become larger and more closely spaced. This indicates that some of the electron's kinetic energy has been diverted from ExB drift into spiral motion. We can check this by finding the average velocity in the x direction at the start and the end of the trajectory:

```
vIn = NIntegrate[x'[t] /. ans1[[1]], {t, 0, 2.×10⁻⁸}] / (2.×10⁻⁸) // Quiet
```

```
-675310.
```

```
vOut = NIntegrate[x'[t + 2.8×10⁻⁷] /. ans1[[1]], {t, 0, 2.×10⁻⁸}] / (2.×10⁻⁸) // Quiet
```

```
492792.
```

```
vIn / vOut
```

```
-1.37038
```

The electron has slowed by nearly 40 percent in the x direction. This would suggest that the erosion rate could be higher on the outboard side of the turnaround compared to the inboard side. This cross corner effect, which is experimentally observed (Fan, Zhou and Gracio 2003), is generally what limits target lifetimes.

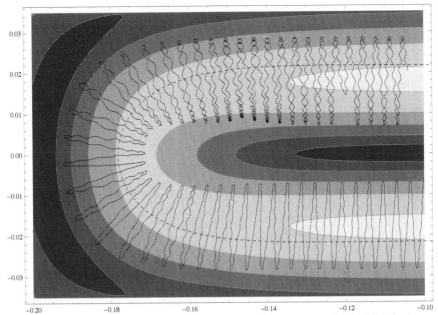

Figure 4- 26. Electron trajectory superimposed on racetrack (dashed line) and contours of the B field parallel to the target.

We can look at a side view of the electron trajectory to get further insights into the effect of the turnaround. Figure 4- 27 shows three planes along the electron path where we examine the trajectory. These are the initial profile, the profile halfway through the turnaround, and the final profile. The initial and final profiles are compared in Figure 4- 28.

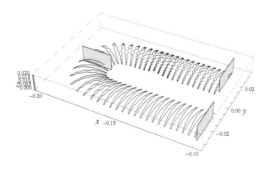

Figure 4- 27. The rectangles indicate where we examine the trajectory in Figure 4- 28 and Figure 4- 29.

The initial profile shows a small amount of motion around the B field lines as noted above. The dashed line shows the profile at the end of the simulation. The orbits are much larger and the width has increased slightly. This would suggest that the erosion groove will be somewhat wider on the outbound side of the turnaround.

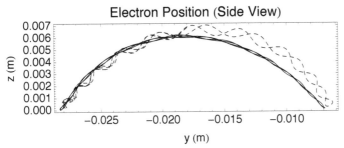

Figure 4- 28. Electron path at the start and end of its trajectory.

A comparison between the initial profile and the midpoint profile is shown in Figure 4- 29. Here the width of the arc has shrunk compared to the incoming profile, suggesting a narrower erosion groove in the turnaround.

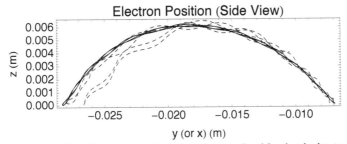

Figure 4- 29. Electron path at the start and midpoint in its trajectory.

From the above results, we can see that even these simple models of electron motion can provide useful insights into the behavior of the magnetron. They allow us to better understand the nature of electron drift as well as the effect of changing field strength on electron motion. However, to go further, we need to include the effect of electron-argon collisions. That is the subject to which we now turn.

5. COLLISIONS

We now have the ability to move electrons over the target surface in an accurate manner. We have seen how they circle the B field lines, arcing backing and forth over the target. Superimposed on top of this is a drift velocity that causes the electrons to circulate over the target. While all of this is good, we are still missing a key ingredient: electron collisions with argon atoms. We need these collisions to generate the ions, which accelerate to the target and cause sputtering. In this chapter, we explore collision processes.

5.1 COLLISION TYPES

In a plasma there are many types of collisions. Electron-argon collisions can be elastic, ionizing, or an excitation. There can be charge exchange collisions, like Ar + Ar⁺ → Ar⁺ + Ar or ion-atom ionizations events like Ar⁺ + Ar → Ar⁺ + Ar⁺ + e⁻. In a full plasma simulation, you would want to include all of these events. For PVD sputtering, most of these processes are unimportant and can be ignored. The processes that are important are those that create Ar ions (needed for sputtering) and those that deplete the electron of energy or change its direction. These are the elastic, excitation and ionization collisions between e and Ar (Sheridan, Goeckner and Goree 1990):

$$e\text{-} + Ar \rightarrow e\text{-} + Ar \quad \text{(elastic, ~1 eV lost)} \quad\quad (5\text{-}1)$$

$$e\text{-} + Ar \rightarrow e\text{-} + Ar^* \quad \text{(excitation, ~12 eV lost)} \quad\quad (5\text{-}2)$$

$$e\text{-} + Ar \rightarrow e\text{-} + Ar\text{+} + e\text{-} \quad \text{(ionization, ~16 eV lost)} \quad (5\text{-}3)$$

Because this is a weakly ionized plasma, collisions between electrons and argon ions or excited argons are too rare to be worth tracking.

5.2 PROBABILITY OF COLLISION

Imagine we have a beam of A particles moving to the right where they impinge on a group of stationary B particles. How many incident particles will collide with the B particles in some thin slice dx? Intuitively we would expect this collision to be proportional to the concentration of A particles and B particles. Thus, we might expect an expression of the form

$$dn_A = -\sigma \, n_A \, n_B \, dx \quad\quad (5\text{-}4)$$

where n_A and n_B are the concentrations of the A and B particles and σ is a proportionality constant. dn_A is the number of A particles per unit volume that interact. The proportionality constant σ has units of area and is called the scattering cross section.

If we multiply both sides of equation (5- 4) by the beam velocity, we can write this in terms of fluxes, $\Gamma = n \, v$:

$$d\Gamma_A = -\sigma \, \Gamma_A \, n_B \, dx \quad\quad (5\text{-}5)$$

We can integrate this to get an expression for the flux of material that does not experience a collision:

$$\Gamma(x) = \Gamma_0 \, exp(-n_B \sigma \, x) \quad\quad (5\text{-}6)$$

In our case we are interested in knowing the probability that an electron will collide with an argon atom. From equation (5- 6) we can see that the probability of not colliding is

$$p_{not}(x) = \frac{\Gamma(x)}{\Gamma_0} = exp(-n_g \, \sigma \, x) \quad\quad (5\text{-}7)$$

Thus the probability of a collision will be

$$p(x) = 1 - p_{not}(x) = 1 - exp(-n_g \, \sigma \, x) \qquad (5\text{-}8)$$

As one traces the trajectory of an electron through the PVD chamber, this expression can be used to determine if a collision has occurred.

A 200 eV electron has an ionization cross section of about 2.4 x 10^{-20} m^2. Using equation (5- 8), we can plot the probability that the electron ionizes an argon atom as a function of distance. First we need the density of the gas, which we can get from the ideal gas law

$$PV = nRT \;\rightarrow\; n_g = \frac{n}{V} = \frac{P}{RT} \qquad (5\text{-}9)$$

At 5 mT and 20C, ng is

$$ng = \frac{0.005 \, torr \; 133.32 \, Pa \, / \, torr}{\frac{8.314 \, Pa \, m^3}{mol \, K} \; 293 \, K}$$

$$\frac{0.000273645 \, mol}{m^3}$$

Because the universal gas constant R is typically in SI units, we need to convert pressure and temperature to Pascal and Kelvin. We actually need the number density of atoms not the molar density. The above value needs to be multiplied by Avogadro's number:

$$ng = ng * 6.022 \times 10^{23} \; \frac{atoms}{mol}$$

$$\frac{1.64789 \times 10^{20} \, atoms}{m^3}$$

With the density of the gas determined, we can go ahead and plot the ionization probability.

```
Plot[1 - Exp[- 1.65 10²⁰ 2.4 × 10⁻²⁰ x], {x, 0, 1},
 Frame -> True, Axes → None,
 BaseStyle → {14, FontFamily → "Helvetica"},
 FrameLabel -> {"Distance (m)", "Probability of Collision"},
 PlotLabel → "Electron-Argon Ionization Probability",
 PlotPoints → 1000,
 PlotStyle → {Black},
 ImageSize → Large, PlotRange → All,
 Epilog → Text["5mT, 20°C", {0.8, 0.1}]]
```

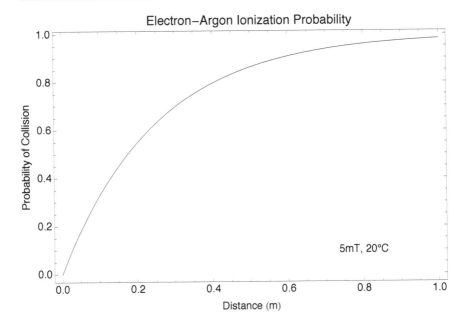

Figure 5- 1. Probability of ionization for a 200 eV electron.

This plot highlights the need for the magnetron in PVD sputtering. Without a magnetic field to confine the electrons to the vicinity of the target, most electrons would fly off to the chamber walls before ionizing a single atom.

5.3 CROSS SECTIONS

The cross sections σ for the three collisions of interest have been experimentally measured by various groups. A readily available source of electron-argon cross section data comes from Hayashi (Hayashi 2005).

(Another source of cross sections is Yanguas-Gil, et al. (Yanguas-Gil, Cotrino and Alves 2005).) The cross sections are a function of energy, so the data are tabulated over the electron energies of interest. In our case, that is from a few eV to a few hundred eV.

The cross sections for e-Ar are listed in Appendix A. They can also be downloaded from www.computationalexplorations.com. Later we will need these cross sections to predict when an electron has collided with an argon atom.

We can easily read the cross section data file into Mathematica. The following commands set Mathematica's current directory to the one where our data file is located. Then we import the file into a variable called data and display the first ten lines of it.

```
SetDirectory["/Users/jack/jProjects/PVDBook-2013/CrossSections"]

/Users/jack/jProjects/PVDBook-2013/CrossSections

data = Import["ArCrossSections.csv"];

Grid[data[[1 ;; 10]]]
```

e-Ar cross sections adapted from Hayashi
Energy in eV, cross
sections in 10^-20 m^2

Energy	Elastic	Excitation	Ionization	Total
0.	7.79	0.	0.	7.79
0.5	0.473	0.	0.	0.473
1.	1.43	0.	0.	1.43
1.5	2.41	0.	0.	2.41
2.	3.52	0.	0.	3.52
2.5	4.53	0.	0.	4.53
3.	5.5	0.	0.	5.5

Since our electron could have any energy, not just the energies listed in the table, we need to interpolate between energies. The next lines do a linear interpolation for each of the four cross sections, creating four interpolation functions. These functions can be used like any other Mathematica functions. They can be plotted, differentiated and so on. The Mathematica code makes use of the Map function to select only those portions of the data table that are of interest. For elastic collisions, we only need the first two columns of data, so Map can be used to extract those. Because the file stores the data in units of $10^{-20}m^2$, we need to multiply by 10^{-20} to work in m^2 units.

```
sigEl = Interpolation[Map[{#[[1]], #[[2]] 10^-16 10^-4} &,
    data[[4 ;; -1]]], InterpolationOrder → 1]
```

```
InterpolatingFunction[{{0., 10000.}}, <>]
```

```
sigEx = Interpolation[Map[{#[[1]], #[[3]] 10^-16 10^-4} &,
    data[[4 ;; -1]]], InterpolationOrder → 1]
```

```
InterpolatingFunction[{{0., 10000.}}, <>]
```

```
sigIon = Interpolation[Map[{#[[1]], #[[4]] 10^-16 10^-4} &,
    data[[4 ;; -1]]], InterpolationOrder → 1]
```

```
InterpolatingFunction[{{0., 10000.}}, <>]
```

```
sigTot = Interpolation[Map[{#[[1]], #[[5]] 10^-16 10^-4} &,
    data[[4 ;; -1]]], InterpolationOrder → 1]
```

```
InterpolatingFunction[{{0., 10000.}}, <>]
```

In Figure 5- 2 these functions are plotted to show the cross sections for elastic, excitation, and ionization collisions.

```
Plot[{sigEl[e], sigEx[e], sigIon[e]}, {e, 0, 1000},
 Frame -> True, Axes → None,
 BaseStyle → {14, FontFamily → "Helvetica"},
 FrameLabel -> {"Energy (eV)", "Cross Section (m^2)"},
 PlotLabel → "Cross Sections",
 PlotPoints → 1000,
 ImageSize → Large, PlotRange → All,
 PlotLegends → Placed[LineLegend[{"Elastic", "Excitation", "Ionization"},
    LabelStyle → {FontFamily → "Helvetica", 14}], {0.7, .8}]
]
```

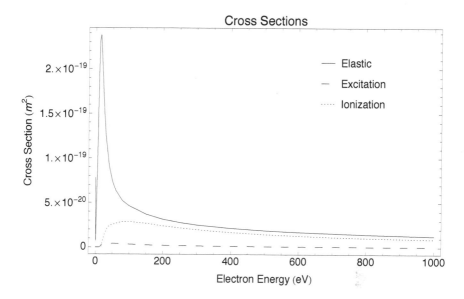

Figure 5- 2. Electron – argon cross sections for three different reactions. The excitation curve comes from summing the results of the first five excitations reported by (Hayashi 2005)

In typical PVD systems, the cathode voltage is in the 300 to 500 volt range. That sets an upper limit on the electron energy. We see that over this energy range elastic collisions are the most common, followed by ionization and then excitation collisions. This can be clearly seen in Figure 5- 3 where the probability of collision is plotted versus distance for a 100 eV electron. We can see that the electron must travel over a meter before it has a 50-50 chance of having an excitation type collision with an argon atom. In contrast, it has a 50 percent probability of having an ionizing collision after 13.8 cm.

The collision probability is heavily influenced by the electron energy, which effects the cross section and the pressure, which effects the number density. These effects are seen in Figure 5- 4 and Figure 5- 5. Because the cross section for ionization falls for energies above 90 eV, high energy electrons must travel farther on average to ionize than those of lower energy. Ionization can be greatly boosted by increasing the pressure. The distance traveled before a collision drops inversely with pressure (Figure 5- 5).

Code 5- 1. Functions for probability of different collision types

```
probEl[Ptorr_, Tkelvin_, energy_, x_] :=
```
$$1 - \text{Exp}\left[-\frac{\text{Ptorr } 133.32}{8.314 \text{ Tkelvin}}\ 6.022 \times 10^{23}\ \text{sigEl[energy] } x\right]$$

```
probEx[Ptorr_, Tkelvin_, energy_, x_] :=
```
$$1 - \text{Exp}\left[-\frac{\text{Ptorr } 133.32}{8.314 \text{ Tkelvin}}\ 6.022 \times 10^{23}\ \text{sigEx[energy] } x\right]$$

```
probIon[Ptorr_, Tkelvin_, energy_, x_] :=
```
$$1 - \text{Exp}\left[-\frac{\text{Ptorr } 133.32}{8.314 \text{ Tkelvin}}\ 6.022 \times 10^{23}\ \text{sigIon[energy] } x\right]$$

```
Plot[{probEl[.005, 273, 100, x],
  probEx[.005, 273, 100, x],
  probIon[.005, 273, 100, x]}, {x, 0, 1},
Frame -> True, Axes → None,
BaseStyle → {14, FontFamily → "Helvetica"},
FrameLabel -> {"Distance (m)", "Probability of Collision"},
PlotLabel → "Electron-Argon Collision Probability",
PlotPoints → 1000,
PlotStyle → {Black, {Black, Dashing[Large]}, {Black, Dashing[Tiny]}},
ImageSize → Large, PlotRange → All,
PlotLegends → Placed[LineLegend[{"Elastic", "Excitation", "Ionization"},
  LabelStyle → {FontFamily → "Helvetica", 14}], {0.7, .7}]]
```

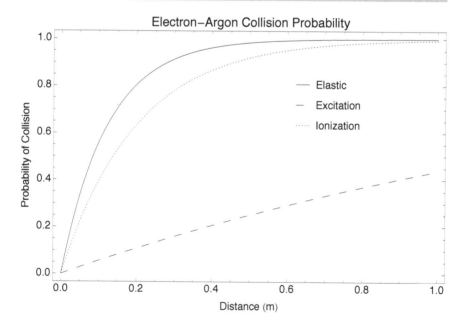

Figure 5- 3. Probability that an electron will undergo an elastic, excitation or ionizing collision with an argon atom, as a function of distance traveled.

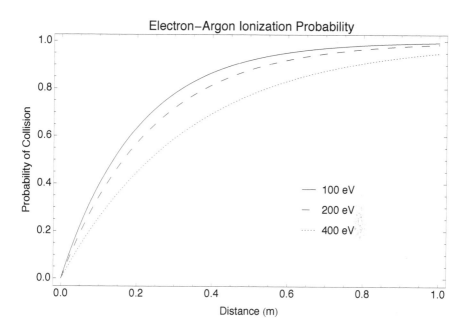

Figure 5- 4. Effect of electron energy on ionization probability. P=5mT, T=20C

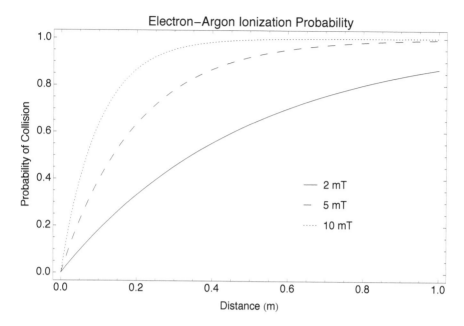

Figure 5- 5. Effect of gas pressure on ionization probability. E=100eV, T=20C

5.4 ENERGY CHANGE

After colliding with an argon atom, the electron loses some energy. In an elastic collision, the energy loss is very small due to the huge difference in masses between the electron and the argon atom. The loss is a fraction of an electron volt. For an excitation collision, the loss varies depending on what excited state the argon moves into. The typical loss is about 12 eV.

The minimum energy required to ionize an argon atom is 15.76 eV. The electron can lose more energy than this, however. For instance, when the argon atom is ionized, it ejects an electron. This electron leaves the atom with some energy, typically several eV. If this electron is ejected with 10 eV of energy, then the incoming electron needs to give up 15.76 + 10 = 25.76 eV of energy. Following Buyle (G. Buyle 2005), we will use the 10 eV value when we start sputtering in the next chapter.

5.5 VELOCITY CHANGE

When we simulate electron trajectories that include collisions, we need to know how the velocity changes after each collision event. The electron velocity, being a vector, has both a magnitude and a direction. For a nonrelativistic electron, the magnitude of the velocity is easily calculated from its kinetic energy:

$$v = \sqrt{\frac{2E}{m}} \qquad (5\text{-}10)$$

For instance, a 100eV electron is moving with a speed of

$$v = \sqrt{\frac{2 \times 100 \text{ eV } 1.602 \times 10^{-19} \text{ J} / \text{eV}}{9.11 \times 10^{-31} \text{ kg}}} \quad / . \text{ J} \to \text{kg m}^2 / \text{s}^2$$

$$5.93044 \times 10^6 \sqrt{\frac{\text{m}^2}{\text{s}^2}}$$

After a collision, we can determine the new velocity magnitude by simply substituting the new energy into equation (5-10) above.

The change in direction is a bit trickier. Consider a simplified case in which the electron is traveling in the z direction and it strikes an argon at the origin of our coordinate system. After the collision, the electron is moving in a new direction, specified by angles ψ and χ (Figure 5-6).

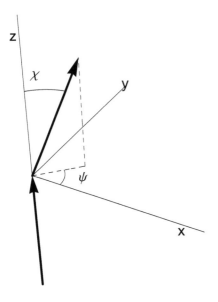

Figure 5- 6. An electron moving vertically strikes an Ar atom at the origin and is deflected by ψ, χ degrees.

Because of the symmetry of the problem, it can be scattered in any azimuthal direction with equal probability. We can simulate this using a random number generator.

$$\psi = 2\pi\, rand()\qquad\qquad(5\text{-}11)$$

The axial deflection is more difficult and depends on the energy of the electron. High-energy electrons are unlikely to scatter by a large angle. An expression for determining the axial scatter angle has been developed by Okhrimovskyy et al. (Okhrimovskyy , Bogaerts and Gijbels 2002) and is

$$\cos(\chi) = 1 - \frac{2r}{1 + 8\epsilon(1-r)}\qquad\qquad(5\text{-}12)$$

where r is a random number between 0 and 1 and ε is the dimensionless electron energy E/E_0 where E_0 is 27.21 eV, the atomic unit of energy. The effect of electron energy on the distribution of angles is shown in Figure 5- 7. As expected, high-energy electrons show less deflection than low-energy electrons.

```
h1 = Histogram[
   {Map[ArcCos[1 - (2 #)/(1 + 8 (50 / 27.21) (1 - #))] 180/π &, RandomReal[{0, 1}, 100 000]],
    Map[ArcCos[1 - (2 #)/(1 + 8 (500 / 27.21) (1 - #))] 180/π &, RandomReal[{0, 1}, 100 000]]},
   200,
   Frame -> True, Axes → None,
   BaseStyle → {14, FontFamily → "Helvetica"},
   FrameLabel -> {"Angle (degrees)", "Number"},
   ChartStyle → {GrayLevel[.5], GrayLevel[.2]},
   Epilog → {
     Text["50 eV", {50, 2000}],
     Text["500 eV", {25, 5000}]},
   ImageSize → Large
]
```

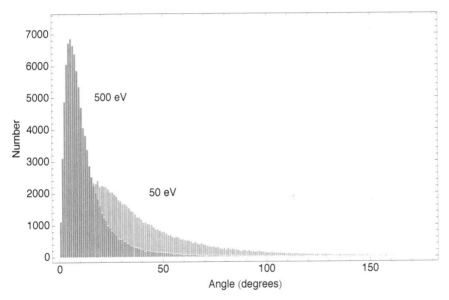

Figure 5- 7. Distribution of axial deflection χ, for two electron energies.

From basic trigonometry, we can see that an electron that was traveling in the z direction is now traveling in a new direction given by

$$\begin{bmatrix} 0 \\ 0 \\ 1 \end{bmatrix} v_{mag} \rightarrow \begin{bmatrix} \sin(\chi)\cos(\psi) \\ \sin(\chi)\sin(\psi) \\ \cos(\chi) \end{bmatrix} v_{mag}'$$

where v_{mag} is the starting velocity magnitude and v_{mag}' is the magnitude after the collision. With this expression, we can calculate the x, y and z components of the velocity after a collision.

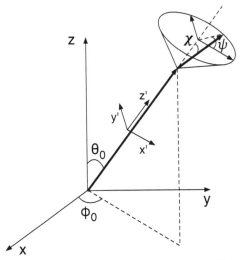

Figure 5- 8. Angle definitions for coordinate transformations.

For the general case, where the electron is moving in an arbitrary direction, we first need to rotate the coordinate system so the electron is moving in the positive z direction. We then calculate the deflection after collision as before and then rotate back to the original coordinate system. As shown in Figure 5-8, an electron will initially be moving in some direction defined by the angles θ_0 and ϕ_0. These can be calculated from the Cartesian coordinates like so:

$$\theta_0 = arccos\left(\frac{v_z}{v_{mag}}\right) \tag{5- 13}$$

$$\phi_0 = arctan\left(\frac{v_y}{v_x}\right) \tag{5- 14}$$

where $v_{mag} = \sqrt{v_x^2 + v_y^2 + v_z^2}$. To transform back to our original coordinate system we first rotate about the z axis by θ_0 and then about the y axis by ϕ_0. Mathematica can easily generate these rotations. They can then be combined with the dot product:

```
m1 = RotationMatrix[φ₀, {0, 0, 1}]
```

$$\{\{\text{Cos}[\phi_0], -\text{Sin}[\phi_0], 0\}, \{\text{Sin}[\phi_0], \text{Cos}[\phi_0], 0\}, \{0, 0, 1\}\}$$

```
m2 = RotationMatrix[θ₀, {0, 1, 0}]
```

$$\{\{\text{Cos}[\theta_0], 0, \text{Sin}[\theta_0]\}, \{0, 1, 0\}, \{-\text{Sin}[\theta_0], 0, \text{Cos}[\theta_0]\}\}$$

```
m1.m2 // MatrixForm
```

$$\begin{pmatrix} \text{Cos}[\theta_0]\,\text{Cos}[\phi_0] & -\text{Sin}[\phi_0] & \text{Cos}[\phi_0]\,\text{Sin}[\theta_0] \\ \text{Cos}[\theta_0]\,\text{Sin}[\phi_0] & \text{Cos}[\phi_0] & \text{Sin}[\theta_0]\,\text{Sin}[\phi_0] \\ -\text{Sin}[\theta_0] & 0 & \text{Cos}[\theta_0] \end{pmatrix}$$

Once we calculate the new velocity magnitude ψ and χ we can get the velocity components in our lab system through this matrix equation:

$$\begin{bmatrix} v_{x1} \\ v_{y1} \\ v_{z1} \end{bmatrix} = \begin{bmatrix} \cos(\theta_0)\cos(\phi_0) & -\sin(\phi_0) & \sin(\theta_0)\cos(\phi_0) \\ \cos(\theta_0)\sin(\phi_0) & \cos(\phi_0) & \sin(\theta_0)\sin(\phi_0) \\ -\sin(\theta_0) & 0 & \cos(\theta_0) \end{bmatrix} \begin{bmatrix} \sin(\chi)\cos(\psi) \\ \sin(\chi)\sin(\psi) \\ \cos(\chi) \end{bmatrix} v_{mag}'$$

To summarize, the steps required to calculate the new electron velocity after a collisions are as follows:

1. Calculate angles θ_0 and ϕ_0 using equations (5- 13) and (5- 14).
2. Calculate the new electron velocity magnitude by subtracting the energy lost through the collision and applying equation (5- 10)
3. Calculate the deflection angles χ and ψ using equations (5- 11) and (5- 12).
4. Apply the above matrix equation to get new velocity components in our lab system

A Mathematica function that does this transformation is shown in Code 5- 2.

Code 5- 2. Calculate new electron velocity after a collision.

```
newVel[{vx_,vy_,vz_}]:=Module[{vmag,vmagNew,θ0,φ0,r,ψ,χ,ener
gy,energyNew},
  vmag=Norm[{vx,vy,vz}];
  energy=0.5 9.11 10⁻³¹ vmag²/(1.602 10⁻¹⁹);
  energyNew =energy-12;
```

```
Print[vmag," energy = ",energy, " ", energyNew];
vmagNew=;
θ0=ArcCos[vz/vmag];
φ0=ArcTan[vx,vy];
r=RandomReal[];
ψ=ArcCos[1-(2 r)/(1+8(energy/27.21)(1-r))];
χ=2π RandomReal[];
Print[θ0," ",φ0, " ",ψ, " χ= ", χ," vmagNew= ",vmagNew];
m1=RotationMatrix[φ0,{0,0,1}];
 m2=RotationMatrix[θ0,{0,1,0}];
Return[m2.m1.{Sin[χ]Cos[ψ],Sin[χ]Sin[ψ],Cos[χ]}vmagNew];

]
```

5.6 ANOMALOUS BOHM DIFFUSION

In the next chapter, when we add gas collisions to our electron transport model, we will find that the predicted erosion profiles are narrower than actual erosion profiles. High-frequency oscillations in the electric field create what is effectively another type of collision, which scatters the electrons and broadens the erosion profile. This effect, first described by Bohm (Bohm 1949), is called Bohm diffusion. It provides an additional mechanism by which electrons can escape from the magnetic field that confines them.

These scattering events occur with a frequency proportional to the cyclotron frequency (Bultinck, et al. 2010)

$$f_{Bohm} = K_{Bohm} f_{cy} \qquad (5\text{-}15)$$

where $f_{cy} = q B / 2 \pi m$. The proportionality constant K_{Bohm} is experimentally determined and typically found to be about $1/16$ (Chen 1984).

The probability of one of these scattering events occurring during a time interval Δt is

$$p_{Bohm} = 1 - \exp(-\Delta t\, f_{Bohm}) \qquad (5\text{-}16)$$

This scattering is somewhat different than the gas collisions we are used to. During a Bohm collision, the electron doesn't lose energy; instead it scatters in a direction perpendicular to the magnetic field lines. We can implement this in a few steps. First, transform the electron velocity vector to a coordinate system where the z axis is aligned with the local B field. Second, generate a random azimuthal angle ϕ representing the new scattering direction. Third, transform back to the original Cartesian coordinate system. The code below implements this.

Code 5- 3. Function for calculating the effect of Bohm scattering.

```
NewVelBohm[v_, B_] := Module[{sph, m1, m2, vPrime, vPerp, θ, vNew},
  (*Convert B vector to spherical coordinates*)
  sph = CoordinateTransform["Cartesian" → "Spherical", B];
  (*Find matrices for rotating velocity vector about φ, θ*)
  m1 = RotationMatrix[-sph[[3]], {0, 0, 1}];
  m2 = RotationMatrix[-sph[[2]], {0, 1, 0}];
  (*Transform velocity vector*)
  vPrime = m2.m1.v;
  (*Find and randomize perpendicular component*)
  vPerp = Norm[vPrime[[1 ;; 2]]];
  θ = RandomReal[{0, 2 π}];
  vNew = {vPerp Cos[θ], vPerp Sin[θ], vPrime[[3]]};
  (*Transform back to std coordinates*)
  Return[m1ᵀ.m2ᵀ.vNew];
]
```

We can test this by considering a case where the local B field is pointing in the direction $\{1, 1, 1\}$ and the electron has a velocity vector of $\{1, 0, 0\}$. Below we call our function fifty times and plot the new velocity vectors generated.

```
B = {1, 1, 1};
vects = Map[NewVelBohm[{1, 0, 0}, B] &, Range[1, 50]];
```

```
Graphics3D[Join[{Dashed, Arrow[{{0, 0, 0}, B}], Dashing[{}], Black}, Map[
  Arrow[{{0, 0, 0}, #}] &, vects]]
]
```

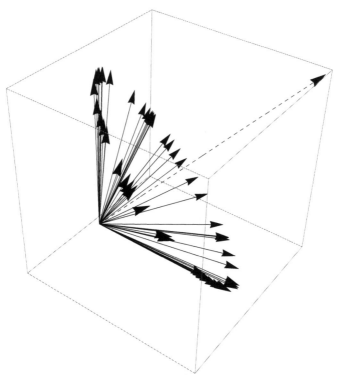

Figure 5- 9. Bohm scattering shifts the electron's velocity vector about the local B field vector, indicated by the dashed line.

As the plot shows, the electron has an equal probability of scattering into any angle about the B field line.

6. EROSION

We can now model the full erosion process. By combining the electron trajectories of Chapter 4 with the electron collisions of Chapter 5, the sputtering of the target can be simulated, atom by atom. By doing this repeatedly, we can generate an erosion profile on the target surface.

6.1 TRAJECTORIES WITH COLLISIONS

In order to model target erosion, we will alter our electron trajectory code and include collisions. The collisions change the energy and direction of the electron, as well as possibly create new ion-electron pairs.

To simulate this, we run our trajectory model for a brief time compared to the mean time between collisions. We then roll the dice and see if a collision took place. If not, run the trajectory model for another short time. If there is a collision, we roll the dice again to determine the type of collision (elastic, excitation or ionization). The appropriate amount of energy is then removed from the election, and the new direction of the electron is determined. If the collision was an ionization, the positions of the new electron and ion are stored in a list. Once the electron no longer has sufficient energy to ionize, we are done with it and can go on to the next electron. In this way, we can process many electrons and generate a large list of ionization events. The locations of those events tell us where the sputtering takes place.

Pseudocode for this procedure is shown in Code 6- 1. We will implement this by creating a few functions in Mathematica. The first one, moveIt, moves the electron through one of those small time steps. This is done by modifying the trajectory code from Chapter 4.

Code 6- 1. Pseudocode for moving a single electron, with collisions
```
While t<tmax and TotalEnery<16 eV
```

```
        Move electron for time delt
        Check for collision
        If collision then
            Find Collision type
            Calculate new velocity
            Record collision position
        End if
        Record electron position and velocity
    Update time step
    End While
```

The code in Code 6- 2. shows the idea. In this function, we specify the starting location and velocity of the electron, as well as the amount of time we want to move it. We then solve the equations of motion using NDSolve as before. When these calculations are done, the new position and velocity are returned.

Code 6- 2. Moving a single electron one time step.

```
moveIt[x0_, v0_, tstep_] := Module[{ans},
  (*Print["MoveIt:v0 = ",v0];*)
  ans = NDSolve[{
    (*Equations of motion*)

    x''[t] == -q/
      m*(y'[t]*Bz[{x[t], y[t], z[t]}, magpack] -
        z'[t]*By[{x[t], y[t], z[t]}, magpack]),
    y''[t] == -q/m*(-x'[t]*Bz[{x[t], y[t], z[t]}, magpack]
+
        z'[t]*Bx[{x[t], y[t], z[t]}, magpack]),
    z''[t] == -q/
      m*(Ez[z[t], -300, 0.001] +
        x'[t]*By[{x[t], y[t], z[t]}, magpack] -
        y'[t]*Bx[{x[t], y[t], z[t]}, magpack]),
    (*Starting position*)
    x[0] == x0[[1]], y[0] == x0[[2]],
    z[0] == x0[[3]],
    (*Starting velocity*)
    x'[0] == v0[[1]], y'[0] == v0[[2]],
    z'[0] == v0[[3]]},
   {x, y, z},
   {t, 0.0, tstep},
   MaxStepSize -> 10^-11];
  (*Return the position and velocity at t=tstep*)

  Return[{{x[tstep], y[tstep], z[tstep]} /. ans[[1]],
```

```
   {x'[tstep], y'[tstep], z'[tstep]} /. ans[[1]]}];
  ];
```

Code 6- 3. Calculate the probability of collision
```
collProb[n_,KE_,delx_]:=1-Exp[-n sigTot[KE] delx]
```

Next, we need to see if a collision has taken place. The function collProb (Code 6- 3) gives us the probability of a collision. By comparing this to a random number, we can see if a collision took place:

```
RandomReal[]<CollProb[N, energy, delx]
```

If a collision took place, we then must determine the type of collision. This is done in Code 6- 4. Using a Which statement, we test a random number against the probability of an elastic collision, σ_{el}/σ_{tot}. If the collision satisfies this criterion and has sufficient kinetic energy, we declare the collision to be elastic. If not, we test against excitation and ionization events in the same way.

Code 6- 4. Decide the type of collision
```
DecideEvent[energy_] := Module[{ans, r, TotalSig, sEl, sEx},
  r = RandomReal[];
  sEl = sigElas[energy];
  sEx = sigExTot[energy];
  TotalSig = sigTot[energy];

  ans = Which[
    r < sEl/TotalSig && energy > 1, 1, (*elastic*)

    r < (sEl + sEx)/TotalSig && energy > 12, 2,
(*excitation*)

    energy > 16, 3, (*ionization*)
    True, 0
    ]
  ]
```

The electron velocity after the collision is determined with the newVel function from Code 5- 2.

Code 6- 5. After a collision, store the location in a list

```
addPtToList[pt_, type_, collList_, eList_] :=
 Module[{r = RandomReal[], temCollList = collList, temEList = eList},
  (*Add pt to collision list*)
  temCollList = Append[temCollList, {pt, type}];

  (*For ionizations add new electron to list*)
  If [type == 3,
   temEList = Append[temEList, pt];
   (*If secondary is emitted, add it to list*)
   If[RandomReal[] < seec,
    temEList = Append[temEList, {pt[[1]], pt[[2]], 0.00001}];
   ];
  ];
  Return[{temCollList, temEList}];
 ]
```

After the new velocity is determined, the location of the collision is stored with the function addPtToList (Code 6- 5). This function may also update a list of electrons. For an elastic or excitation collision the location and type of collision are stored in the list collList. This information is also stored for ionization collisions. However, an ionization will produce an argon ion, which will sputter the target surface. When the ion strikes the target there is some chance that a secondary electron will be emitted. These secondaries sustain the plasma by accelerating through the sheath and generating more electron-ion pairs. To see if a secondary electron is produced, we compare a random number to the secondary electron emission coefficient, seec. This coefficient typically ranges between 0.01 and 0.1 depending on the target material and electron energy (Liberman and Lichtenberg 2005). If a secondary is produced, we add it to the list of active electrons. We also add the electron liberated by the ionization. This latter electron will only factor into sputtering provided it is born in the sheath so that it can accumulate sufficient energy to generate ions.

These functions are all called by a higher-level function runE, which follows one electron until its energy falls below the ionization energy (Code 6- 6) . This function returns four lists: the electron position at each time, the velocity at each time, a list of collision locations, and a list of locations where electrons were generated.

Code 6- 6. Main function for moving a single electron, with collisions
```
runE[p0_, v0_, tmax_, tstep_] :=
 Module[{time, p1, p2, v1, v2, delx, KE, prob, r, type = 0,
vMag,
   vMag2, plist = {p0}, vlist = {v0}, eList = {}, collList =
{}},
  time = 0.0;
  p1 = p0; v1 = v0;
```

```
While[time < tmax && totEnergy[p1, v1] > 26.,
  {p2, v2} = moveIt[p1, v1, tstep];

  delx = EuclideanDistance[p1, p2];
  vMag = Norm[v2];
  KE = 0.5 m vMag^2 1 (*eV*)/1.602 10^-19(*J*);
  prob = collProb[nGas, KE, delx];
  r = RandomReal[];
  If[ r < prob,
   type = decideEvent[KE];
   v2 = newVel[v2, KE, type];(*Check this!!!*)
   {collList,
     eList} = addPtToList[p2, type, collList, eList];
   ];
  (*Update position and velocity lists for time step*)

  plist = Append[plist, p2];
  vlist = Append[vlist, v2];

  (*Update time and position*)
  time = time + tstep;
  p1 = p2; v1 = v2;
  ];(*while loop*)
 Return[{plist, vlist, collList, eList}];
 ]
```

With the code in place, we can now run an electron through its paces and see what collisions it undergoes.

6.2 SINGLE ELECTRON

We will move our electrons around a magnet array defined by

```
magpack2={
{{-8cm,-1cm,-2cm},{16cm,2cm,1cm}, Br},(*center mags*)
{{-10cm,-2cm,-2cm},{1cm,4cm,1cm}, -Br},(*bottom row*)
{{-10cm,2cm,-2cm},{20cm,1cm,1cm}, -Br},(*right side*)
{{9cm,-2cm,-2cm},{1cm,4cm,1cm}, -Br},(*top row*)
{{-10cm,-3cm,-2cm},{20cm,1cm,1cm}, -Br}}(*left side*)
```

The discharge voltage will be -300 V with an assumed sheath thickness of 1 mm. The target surface is located at z = 0, one centimeter above the magnet array.

Using the program runE, we can model a full electron trajectory with this line:

```
(ans2 = runE[{0., 0.015, 0.0}, {0., 0., 0.}, 5 10^-8, 3. 10^-
11]; //
   Timing)
```

On my 2012 MacBook Air it takes about four minutes to simulate the electron moving for 0.05 μs. In the variable ans2, we have stored the position and velocity as a function of time as well as the location of the collisions and any secondary electrons produced. The trajectory can be plotted with this code:

```
Graphics3D[{
   Line[ans2[[1]]],
   White, PointSize[Large],
   Point[Select[ans2[[3]], #[[2]] == 1 &][[All, 1]]],
   Gray, PointSize[Large],
   Point[Select[ans2[[3]], #[[2]] == 2 &][[All, 1]]],
   Black, PointSize[Large],
   Point[Select[ans2[[3]], #[[2]] == 3 &][[All, 1]]]},
  Axes -> True,
  AxesLabel -> {x, y, z},
  BoxRatios -> {5, 1, .3},
  ImageSize -> 600]
```

The Line function instructs Graphics3D to plot the list of positions as a line. The next lines extract the collision locations and depending on the type plot them as white, gray or black dots. The result can be seen in Figure 6- 1.

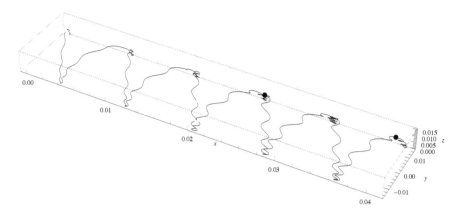

Figure 6- 1. Electron trajectory with collisions.

This trajectory looks very similar to those generated in Chapter 4. In order to fully appreciate the effect of collisions on the trajectory, we need to run for longer times. Figure 6- 2 shows the trajectory where the time was increased to 1.2 μs. We can see that the trajectory is significantly altered by the collisions. Through these collisions, the electron is losing energy. This is shown in Figure 6- 3. Once the electron gets below 26 eV, the run is stopped.

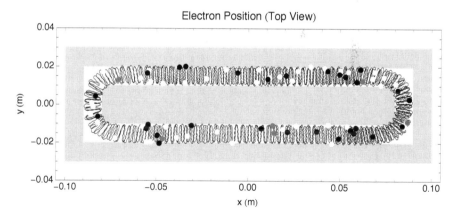

Figure 6- 2. Electron trajectory starting at x = 0, y = 10mm.

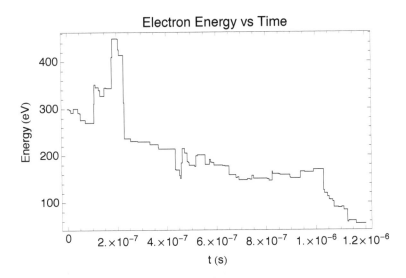

Figure 6- 3. Electron energy decreases through collisions.

The electron trajectory depends strongly on the starting position of the electron. If the electron leaves the target in a region of strong B field, it will have a smaller Larmor radius and a smaller ExB drift velocity. On the other hand, if the electron starts where the field is weak, it will be only weakly bound to the target region resulting in large, sloppy orbits. These can be seen in Figure 6- 4 through Figure 6- 7.

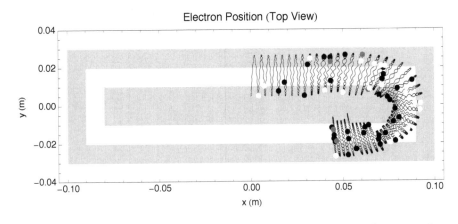

Figure 6- 4. Electron trajectory for a tightly bound electron starting at y = 5 mm.

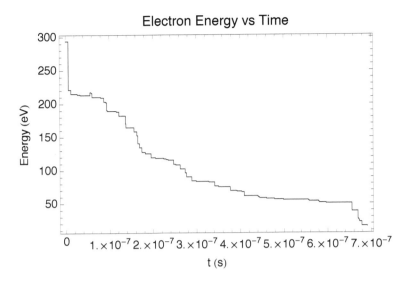

Figure 6- 5. Electron energy for loosely bound electron.

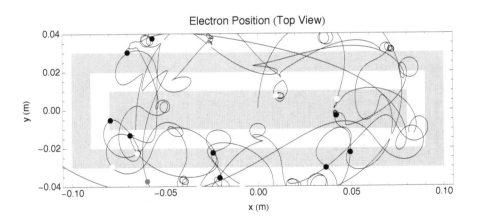

Figure 6- 6. Chaotic trajectory for weakly bound electron, starting at y = 0 mm.

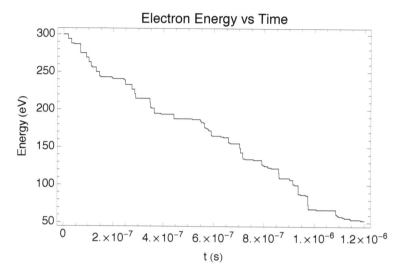

Figure 6- 7. Electron energy decreases through collisions.

6.3 EROSION PROFILE

As the target erodes, a track is worn into its surface. Before modeling the entire erosion profile, we will create a smaller model to reveal the shape of this erosion track. This can be done by running the electrons over a simple magnet array consisting of just two long magnets:

```
magpack = {
    {{-20 cm, -2 cm, -2 cm}, {40 cm, 1 cm, 1 cm}, Br}, (*left mag*)
    {{-20 cm, 1 cm, -2 cm}, {40 cm, 1 cm, 1 cm}, -Br}(*right mag*)
    };
```

Figure 6- 8. Two magnet array.

We will then run 400 electrons along this array and plot the y positions of the sputter events. Before doing this, however, we need to address the issue of computational speed. For all of its flexibility, Mathematica can be very slow. If each electron takes many minutes to run, even a 400 electron set will take a very long time. Fortunately, there are ways to speed up Mathematica. The code to do this is described in Appendix C. Basically we replace the routine moveIt with one that swaps NDSolve with a Runge-Kutta solver and uses the Compile function. It runs about 50x faster. Alternatively, one can code these routines in a compiled language like C or Fortran. That would provide an additional speedup.

The routine runCloud (Code 6- 7) implements a multi-electron run. It assumes we have defined a list called bigEList. Initially we populate this list with ten electrons randomly located on the target surface at x = 0, between y = -2 cm and 2 cm.

```
bigEList=Map[{0.,#,0.00001}&,RandomReal[{-2cm,2cm},10]]
```

Additional electrons are added to this list as ionization events occur.

Code 6- 7. Erosion profile in 2D.
```
runCloud[maxE_] := Module[
```

```
  {numElectrons = 0, p, ans},
  While[
    numElectrons < maxE,

    (*Grab electron from front of list*)
    p = bigEList[[1]];
    p[[1]] =
      0.; (*zero out x position so e doesn't leave magpack*)
      \
(*Print["numE= ",numElectrons," p = ",p];*)

    bigEList = Drop[bigEList, 1];

    (*Launch e and record results*)

    ans = runE[p, {0., 0., 0.}, 2 10^-7, 3. 10^-11];
    bigArList = Join[bigArList, getPts[ans, 3]];
    bigEList = Join[bigEList, ans[[4]]];
    numElectrons = numElectrons + 1;
    myNumE = numElectrons;  (*for monitoring progress*)
    ];
  ];
```

The 400 electron run can be initiated with this line

```
runCloud[400]//Timing
```

The list bigArList holds all of the positions of the argon atoms that are ionized during this run. Because each electron has sufficient energy to ionize several atoms, about 560 ionization events occur. We can plot them in a histogram to get a rough idea of the racetrack shape:

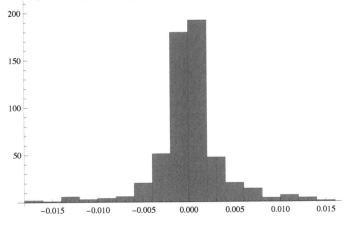

Figure 6- 9. Histogram of erosion rates.

Alternatively we can bin the data with the function BinCounts and generate a line plot, as shown in Figure 6- 10. The y axis has been inverted and subtracted from the target (in gray) to emulate an erosion profile.

```
ListLinePlot[
 Table[{-0.0105 + i*.002, -counts[[i]]/49500}, {i, 1,
Length[counts]}],
 BaseStyle -> {Medium, FontFamily -> "Helvetica"},
 Frame -> True, Axes -> None,
 FrameLabel -> {"y (m)", "z (m)"},
 PlotLabel -> "",
 Filling -> Axis, FillingStyle -> White,
 Prolog -> {Prepend [
    Map[Rectangle[#[[1, 2 ;; 3]], #[[1, 2 ;; 3]] + #[[2, 2 ;;
3]]] &,
     magpack2], GrayLevel[0.3]], {GrayLevel[0.5],
    Rectangle[{-3 cm, -0.5 cm}, {3 cm, 0}]}
   },
 PlotRange -> {{-2.1 cm, 2.1 cm}, {-2.2 cm, 1 cm}},
 AspectRatio -> 0.75, ImageSize -> 400]
```

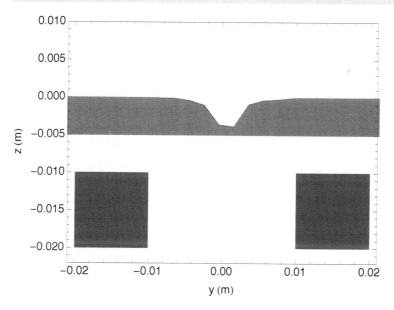

Figure 6- 10. Simulated erosion profile.

We see that the racetrack for this magnetic array is only a few millimeters wide. This would suggest very poor target utilization for this design. However, it is important to note that these Monte Carlo models tend to under-predict the width of the racetrack (Musschoot, et al. 2008). It is thought that anomalous diffusion (Section 5.6) must be included to broaden the racetrack width by providing additional scattering to the electrons. We have not included anomalous diffusion in our erosion modeling, but it could be added using Code 5- 3.

The location of the ionization events is shown in Figure 6- 11. While ionizations occur throughout the region above the magnets, the vast majority of the ionizations occur close to the target surface, centered about y = 0.

```
ListPlot[bigArList[[All, 2 ;; 3]],
  BaseStyle -> {Medium, FontFamily -> "Helvetica"},
  Frame -> True, Axes -> None,
  FrameLabel -> {"y (m)", "z (m)"},
  PlotLabel -> "Ionization Events",
  PlotStyle -> Black,
  Prolog -> {Prepend [
    Map[Rectangle[#[[1, 2 ;; 3]], #[[1, 2 ;; 3]] + #[[2, 2 ;;
3]]] &,
      magpack2], GrayLevel[0.3]], {GrayLevel[0.5],
      Rectangle[{-3 cm, -0.5 cm}, {3 cm, 0}]}
    },
  PlotRange -> {{-2.1 cm, 2.1 cm}, {-2.2 cm, 1 cm}}, ImageSize
-> 400]
```

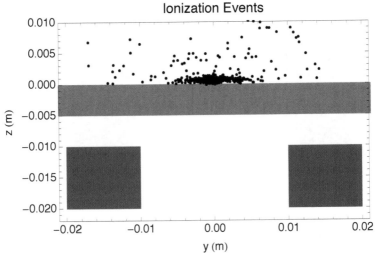

Figure 6- 11. Location of ionization events above target.

6.4 FULL TARGET EROSION

To generate an erosion profile for the entire target, we need to run many hundreds or even thousands of electrons around the racetrack. To do this, we record where on the target the Ar ions fall using a 2D array. The five magnet array we used in section 6.2 is 20 cm by 6 cm. To cover that area with 1mm resolution, we need an array of 201 x 61 = 12,261 elements. Such an array can be created and initialized to zero with the Array function:

```
targGrid=Array[0&,{201,61}];
```

The routine to calculate the erosion profile, runErosion (Code 6- 8), is similar to runCloud. However some care is needed in managing the bigEList list of electrons. It is easy to either run out of electrons or create a list so huge that it consumes vast amounts of memory. To maintain a reasonable list size, we dynamically adjust the secondary electron emission coefficient, seec. When the list is getting small, we boost the coefficient, and when the list is getting large, we shrink it.

This code doesn't include electrons generated through ionization. The majority of these will be created outside of the sheath and will never have the energy required to create ions themselves. Still, this is an approximation and in a more complete implementation, we would include them.

After each electron is run, we cycle through all of the argon ions created and add them to the target array.

Code 6- 8. Generate erosion profile.
```
runErosion[maxE_] := Module[
  {numElectrons = 0, p, ans, sputterList, i, j, k, numPts},
  While[
  numElectrons < maxE,

  If[Length[bigEList] < 1,
    (*Electron queue is empty.  End*)
    Break[],
    (*Plenty of electrons,
    so keep going*)
    (*Grab electron from front of list*)

    p = bigEList[[1]];
    bigEList = Drop[bigEList, 1];
```

```
(*Launch e and record results*)

ans = runE[p, {0., 0., 0.}, 2 10^-7, 3. 10^-11];
sputterList = getPts[ans, 3];
numPts = Length[sputterList];


(*Record sputter evens*)
For[k = 1, k <= numPts, k++,
 i = Round[sputterList[[k, 1]]/0.001] + 101;
 j = Round[sputterList[[k, 2]]/0.001] + 31;
 (*Print[i," j= ",j];*)
 If[0 < i <= 201 && 0 < j <= 51,
  targGrid[[i, j]] += 1;];


 (*Adjust SEEQ to give a reasonable size of e list*)

 SEEQ = Min[100/Length[bigEList], 1];
 If [RandomReal[] < SEEQ,
  bigEList =
    Append[bigEList,
     sputterList[[k]] /. {x_, y_, z_} -> {x, y, 0}];
  ];
 ];
 numElectrons = numElectrons + 1;
 myNumE = numElectrons;
 ]    (*end If*)
 ]    (*end While*)
];
```

The results of the simulation are shown in Figure 6- 12 for a 1000 electron run. A total of 7900 argon ions were generated. Even with this small number of sputter events, a clear erosion profile can be seen. Because this is a Monte Caro simulation, there is statistical noise in the result. To reduce this noise, a larger run is required.

This run took about twelve hours on one core of a Core i7 processor. To run the tens of thousands of electrons needed for a smooth profile, it would make sense to implement this in a compiled language, such as C or Fortran.

With the runErosion program, we have created a framework for predicting the target erosion. By running cases with different pressures and field strengths, we can build a deeper understanding of the sputtering process. Such a tool can also be used to improve the magnetron design; however, some care is required. Any model is a simplification of reality, and these simplifications reduce the accuracy of the prediction. The best approach is to make the model as realistic as possible and then validate it against experimental data. In this case, the model could be improved by adding Bohm diffusion to the electron motion and including electrons generated by ionization.

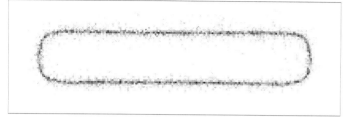

Figure 6- 12. Erosion profile of a full target.

7. DEPOSITION

Once the atoms have been sputtered off the target, they move through the plasma and strike the substrate, building up a film. Often the uniformity of this film is of great practical importance. In this chapter, we will calculate the deposition profile on the substrate using an analogy to radiative heat transfer. This view factor approach is computationally efficient, but it does not include the effect of scattering collisions in the gas phase.

A cross section of our sputter system is shown in Figure 7- 1. The magnetic field is created with the same magnet array we used in earlier chapters. Above is the target. The argon ions sputter atoms off the target surface. These follow a line of sight trajectory to the substrate.

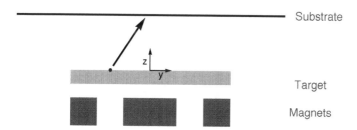

Figure 7- 1. A sputtered atom follows a straight line to the target (ignoring gas collisions).

7.1 RADIATION ANALOGY

To a good approximation, the sputtered atoms leave the target in a cosine distribution. This means that more material is sputtered normal to the surface than at some angle θ. Figure 7- 2 shows the idea. It turns out that when light is emitted or reflected from a rough surface, it behaves in the same way. This is called diffusive reflection in the case of light. There are well-established methods on how to calculate this for light, discussed in many heat transfer books. We can take advantage of these methods for our case.

Figure 7- 2. Material is sputtered in a cosine distribution.

The key concept is that of the view factor or configuration factor. Imagine you are a little person sitting at some arbitrary point on the substrate and you look at the erosion groove of the target. How much of it can you see? Well, that depends on your angle. If it is a very large substrate and you are at the outer reaches of it, the racetrack will appear small and skewed. This is the idea behind the view factor. The formula is given by Siegel and Howell (Siegel and Howell 1992):

$$F_{12} = \frac{1}{A_1} \iint \frac{\cos(\theta_1)\cos(\theta_2)dA_1 dA_2}{\pi r^2} \qquad (7\text{-}1)$$

where F_{12} is the view factor from emitting surface 1 to receiving surface 2. θ_1 is the angle between the normal vector of surface 1 and the line connecting surface 1 and 2. Similarly, θ_2 is the angle between the normal vector of surface 2 and the line connecting 1 and 2. (Figure 7- 3) The distance between the surfaces is r. This is a general formula that can be applied to any shape emitter. The integrals can be solved for specific geometries to generate simpler equations. For our case, we are going to divide the target and substrate into small rectangles. If these rectangles are small enough, then the properties of a single rectangle can be assumed constant over its area. Specifically the flux and angles will be approximately constant over each rectangle. Equation (7- 1) then becomes

$$F_{12} = \frac{\cos(\theta_1)\cos(\theta_2)A_2}{\pi r^2} \qquad (7\text{-}2)$$

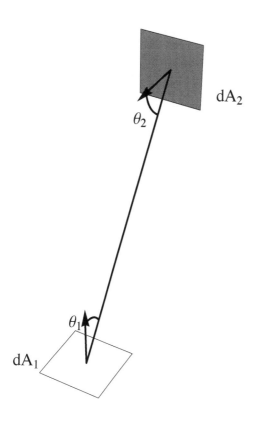

Figure 7- 3. The view factor between surfaces dA_1 and dA_2 depends on angles θ_1 and θ_2.

7.2 DISCRETIZATION

To divide up the substrate into computational cells, we need to make an array of evenly spaced points across the substrate surface. Let's say our substrate is

20 cm wide and 60 cm long. The center point of each cell can be set up using the Table command in Mathematica.

```
cm = 0.01;
gap = 10 cm;   (*targ-substrate distance*)

subMidPts = Flatten[
   Table[
     {j, i, gap},
     {i, -10 cm, 10 cm, 1 cm},
     {j, -30 cm, 30 cm, 1 cm}], 1];
```

This creates an array of points spaced 1 cm apart in x and y. The Flatten command flattens this 2-D matrix into a one-dimensional list of points. Here are the first few points:

```
subMidPts[[1;;11]]

{{-0.3,-0.1,0.1},{-0.29,-0.1,0.1},{-0.28,-0.1,0.1},{-0.27,-
0.1,0.1},{-0.26,-0.1,0.1},{-0.25,-0.1,0.1},{-0.24,-
0.1,0.1},{-0.23,-0.1,0.1},{-0.22,-0.1,0.1},{-0.21,-
0.1,0.1},{-0.2,-0.1,0.1}}
```

We have set the z position of the substrate at 10 cm = 0.1 m.

We also need arrays for the area and normal vector of each cell. Because our cells are all the same size and the substrate is flat, these values are constant for all cells in the array. The normal vectors point in the –z direction.

```
subAreas = Table[1 cm 1 cm, {i, 1, Length[subMidPts]}];

subNorms = Table[{0, 0, -1}, {i, 1, Length[subMidPts]}];
```

Dividing the target into cells is a bit more tricky. Our target is rectangular, like the substrate, but the sputtering only occurs at the racetrack. It is not computationally efficient to include the portions of the target where no sputtering takes place. In addition, the sputter rate varies along the racetrack. It is typically lower in the turnaround regions, for instance. Using the emission model from the previous chapter, we could determine this. However, we saw that it is computationally very expensive to do a full erosion model with sufficient precision in Mathematica. For simplicity, we will

120

assume the emission is uniform around the racetrack. Specifically, we will assume the target emits atoms in a 2 cm wide band around the racetrack.

The coordinates of the racetrack are found by locating the line on the target surface where the B field normal to the surface is zero—that is, where the B field is parallel to the surface. Mathematica can do this for us with a contour plot. Using our function for Bz from Chapter 2, along with the magpack from that chapter, we can plot the line where Bz = 0.

Figure 7- 4. The B-field contour where Bz = 0.

To get the points that make up this contour plot, we must look inside the data structure Mathematica uses to store the contour plot data. This structure is called a Graphics Complex. The Graphics Complex for a contour plot stores all of the points of each contour line, plus additional information needed to generate the plot. We can extract the data points with this command:

```
contourPts = First[Cases[First[cp2] // Normal, Line[pts_] → pts, ∞]];
```

This is a tricky bit of coding is from the StackExchange website (Popkov 2013). It takes the Graphics Complex, named cp2, and first converts it to a list using the Normal command. It then looks through that list for a command of the form Line[pts_] and extracts the points from there.

Using the Length command, we can see how many points are in the list.

```
Length[contourPts]
```

```
455
```

This is probably more points than we need. We can use the Take command to thin this down by four:

```
tPts = Take[contourPts, {1, -1, 4}];
```

```
Length[tPts]
```

These points have only x and y coordinates. We need to add the z component (which is zero). Also, we need to append a copy of the first point to the end of the list to close the loop. This command does both:

```
tPts2 =
Append[Map[{#[[1]],#[[2]],0}&,tPts],{tPts[[1,1]],tPts[[1,2]],
0}];
```

We have now defined the end points of our cells. The midpoints can be found by averaging the two end points of a given cell. This can be done very simply by taking a copy of our list of points and shifting it to the left. We then add this list to the original and divide by two. The last point, which is meaningless, is dropped.

```
targMidPts = 0.5 Drop[tPts2 + RotateLeft[tPts2], -1];
```

To determine the area of each cell we find the distance between the end points and multiply it by the width of our racetrack (2 cm). This can be done with a similar list-shifting approach:

```
targAreas = Map[Norm[#] &, Drop[tPts2 - RotateLeft[tPts2], -
1]*2 cm];
```

Because our target is flat, the normal vectors all point straight up.

```
targNorms = Table[{0, 0, 1}, {i, 1, Length[targMidPts]}];
```

Similarly, the emission rate is assumed constant around the racetrack

```
targEmission = Table[1, {i, 1, Length[targMidPts]}];
```

These arrays could also have been set up using a more conventional procedural approach, with a For loop:

```
For[i = 1, i < Length[tPts2], i++,
  targMidPts[[i]] = 0.5*(tPts2[[i]] + tPts2[[i + 1]]);
  targAreas[[i]] = Norm[tPts2[[i + 1]] - tPts2[[i]]]*2 cm;
  targNorms[[i]] = {0, 0, 1};
  targEmission[[i]] = 1;
  ]
```

7.3 DEPOSITION

The deposition rate at any point on the substrate is the sum over all target cells of the emission rate times the view factor, scaled by the cell areas. We can write this as

$$R_i = \frac{1}{A_i} \sum_{j=1}^{n} E_j F_{ji} A_j \qquad (7\text{-}3)$$

where R_i is the rate at the ith substrate cell and E_j is the emission from the jth target cell. By doing this summation for each point on the substrate, we can determine the deposition rate over the entire surface. To try this, we first need to implement the view factor equation (7-2) as a function:

Code 7- 1. Function for calculating view factors.

```
vfCalc[targIndex_, subIndex_] := Module[{dot1, dot2, diffVect, F12},
  diffVect = targMidPts[[targIndex]] - subMidPts[[subIndex]];
  dot1 = targNorms[[targIndex]].Normalize[diffVect];
  dot2 = subNorms[[subIndex]].Normalize[diffVect];
  F12 = Max[- (dot1 * dot2 * subAreas[[subIndex]]) / (π Norm[diffVect]^2), 0]
]
```

This function takes advantage of the fact that the dot product of any two unit vectors is the cosine of the angle between them. We can now calculate the view factor between any target cell and substrate cell:

```
vfCalc[1,1]
```

5.41812×10^{-6}

The deposition rate at a specific substrate cell can be calculated with this function:

```
depRate[subIndex_] := Module[{cellVFs},
  cellVFs = vfCalc[Range[1, Length[targMidPts]], subIndex];
  Total[cellVFs*targEmission*targAreas]/subAreas[[subIndex]]
]
```

The deposition rate to the first cell of the substrate (in arbitrary units) is

```
depRate[1]
```

```
0.000559759
```

The deposition rate to the entire substrate can be calculated by running this function on all substrate cells:

```
depRates =
  Table[{subMidPts[[i, 1]], subMidPts[[i, 2]], depRate[i]},
{i, 1,
    Length[subMidPts]}];
```

```
depRates[[1 ;; 11]]
```

```
{{-0.3, -0.1, 0.00893494}, {-0.29, -0.1, 0.0102108},
 {-0.28, -0.1, 0.0116781}, {-0.27, -0.1, 0.0133589}, {-0.26, -0.1, 0.0152731},
 {-0.25, -0.1, 0.017437}, {-0.24, -0.1, 0.0198599}, {-0.23, -0.1, 0.0225411},
 {-0.22, -0.1, 0.0254669}, {-0.21, -0.1, 0.028608}, {-0.2, -0.1, 0.0319184}}
```

The depRates list includes the x and y positions of each cell midpoint, in addition to the deposition rate. In this format we can easily generate a contour plot of the deposition on the substrate.

```
ListContourPlot[depRates,
  BaseStyle -> {Medium, FontFamily -> "Helvetica"},
  AspectRatio -> .3, PlotRange -> All, ContourLabels -> All,
  FrameLabel -> {"x (m)", "y (m)"},
  ColorFunction -> ColorData["GrayTones"], ImageSize -> Large]
```

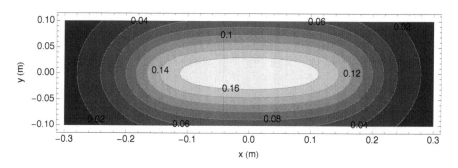

Figure 7- 5. Deposition rate on the substrate in arbitrary units.

The deposition is highest in the center and drops off toward the edges of the substrate. In addition, a significant fraction of the emission misses the substrate.

7.4 UNIFORMITY

From Figure 7- 3 we can see that the uniformity of deposition is poor on this substrate. The deposition rate varies from 0.16 at the center down to 0.02 at the edges. The non-uniformity is both in the x direction and the y direction. One way to partially address the non-uniformity is to put the substrate in motion. For instance, we can picture the substrate on a conveyor belt moving past the target. This will average out the non-uniformity in one direction.

We can simulate this movement by summing the deposition rates in one direction. To do this we first have to convert our one dimensional list of deposition rates into a 2-D array. This can be done with the Split function. This splits the list into sublists at the end of each row. Our criterion is to split the list at any point where the x position of the n and n+1 points differ.

```
depRateArray = Split[Sort[depRates, #2[[1]] > #1[[1]] &], #2[[1]] == #1[[1]] &];
```

Now we sum up each row. When we use the Total function, we are summing up the x and y *positions* as well. Since we are summing over y, we can recover the x position by taking the sum of the x's and dividing by the number of items in the row. Below we use these x positions in our plot.

$$
\text{xRate = Map}\left[\frac{\text{Total[\#]}}{\{\text{Length[depRateArray[[1]]]}, 1, 1\}} \&, \text{depRateArray}\right];
$$

By moving the substrate, we have eliminated all of the variation in y. The only variation remains in x, which we plot below.

```
ListLinePlot[xRate[[All, {1, 3}]],
 Frame -> True,
 Axes → None,
 BaseStyle → {Medium, FontFamily → "Helvetica"},
 PlotLabel → "Deposition Rate in X Direction",
 PlotStyle → Black,
 FrameLabel → {"x (m)", "Deposition Rate (arb.)"},
 ImageSize → Large]
```

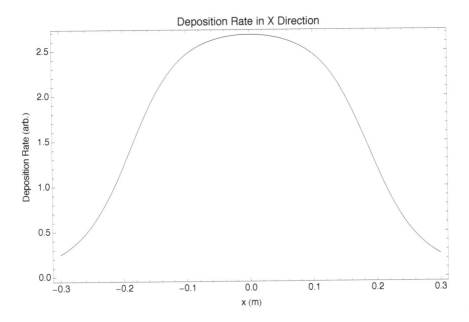

Figure 7- 6. Deposition rate, averaged across the y direction

Of course the non-uniformity is still quite high in the x direction. One way to improve it is to move the substrate closer to the target. Up until now we have been using a gap of 10 cm. We can reduce it to say 5 cm and 2 cm and look at the effect on deposition. Contour plots of the closer spacings are shown in Figure 7- 7. The effect of averaging the deposition in the y direction is shown in Figure 7- 8. The profile goes from being a broad, round hill at 10 cm to a narrow flat plateau at 2 cm. At that spacing, a substrate of 30 cm in length would experience a very uniform deposition. It would also see a higher deposition rate.

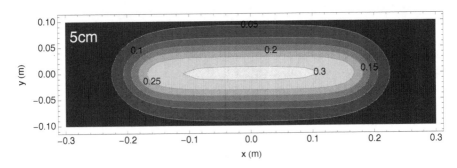

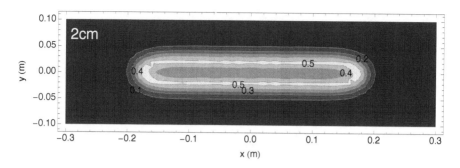

Figure 7- 7. Deposition rates at 5 cm and 2 cm substate-target distances.

Notice the noise in the 2 cm profile. At this spacing we are beginning to violate the assumption we made in deriving equation (7- 2). With our cell size of 1 cm and a gap of only 2 cm, the flux to a given cell can no longer be considered constant over that cell area. At this spacing, the cell size for the substrate cells should be reduced to 0.5 cm or even smaller and the model should be rerun.

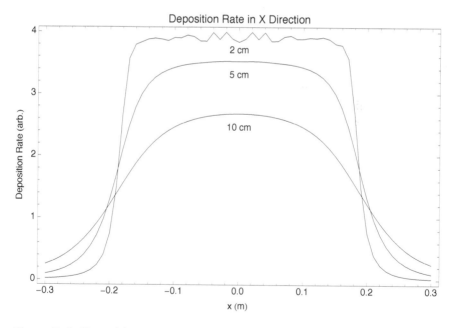

Figure 7- 8. Deposition rate, averaged across the y direction.

This approach to estimating the deposition rate has some limitations. We assumed the sputter rate was uniform around the racetrack. In reality this rate will vary significantly. This could be accounted for using an erosion model, as described in Chapter 6. Even better would be to measure the erosion profile of a spent target and use that data as an input to the model.

A more fundamental limitation of this approach is that we have not accounted for gas collisions. We have assumed that the sputtered atoms follow a straight-line trajectory from target to substrate. At low pressures and small spacing, most atoms will do this. However, some fraction will collide with an argon atom. These collisions alter the direction of the sputtered atom, as well as its energy. Some atoms will even return to the target. The effect of these collisions is to broaden the deposition profile on the substrate. The view factor approach is a surface-to-surface method that ignores gas collisions. In order to include them, we would need an alternative method. The Monte Carlo method is frequently used. The idea is similar to the approach of the last chapter. Sputtered atoms leave the surface with their direction and energy chosen randomly. We roll the dice to see if a collision takes place. If it does, we calculate the change in energy and direction. In this way we can fully account for the physics of the problem. The downside is that these simulations can require considerable computing time. For many cases, the simpler approach taken here is sufficiently accurate.

In contrast to erosion modeling, the physics of deposition modeling is fairly straightforward. Even ignoring gas collisions, accurate models can be easily generated. These can be very helpful in exploring different strategies for improving uniformity. It is generally much faster to change the sputtering system on the computer than in the lab.

EPILOGUE

This is the end of our journey. We have spent quite a bit of time following electrons in our effort to better understand magnetron sputtering. Along the way, we studied the forces acting on the electron, through the magnetic and electric fields. In Chapter 4 we used this knowledge to solve the equations of motion for an electron and trace out its trajectory over the target surface. These electrons first entered the picture when emitted from the target during ion bombardment. We then saw how the local electric field accelerates them across the sheath and into the plasma. The magnetic field of the magnetron then bends the electron trajectory into an orbit around the B field line. This eventually returns the electron to the target surface. To this picture we added electron-atom collisions in chapters 5 and 6, allowing us to model the sputtering process and study target erosion. In the final chapter we have seen how to model the deposition pattern of the sputtered atoms.

Physical vapor deposition is like a huge national park: vast and beautiful. We have taken one path through this park and have seen some amazing sites. But there is much more to explore. Fortunately, many excellent guides are available to show the way. A good overview of thin film processing can be found in:

Donald M. Mattox, *Handbook of Physical Vapor Deposition (PVD) Processing*, 2nd edition, Elsevier, 2010.

For those interested in the aspects of sputtering relevant to semiconductor fabrication, the book by S Rossnagel and my esteemed colleague Ron Powell is worth studying:

R. A. Powell and S. M. Rossnagel, *PVD for Microelectronics: Sputter Deposition Applied to Semiconductor Manufacturing*, Thin Films, volume 26, Academic Press, 1999.

Reactive sputtering is an important method for depositing compounds. A good introduction can be found in:

D. Depla and S. Mahius (eds), *Reactive Sputter Deposition*, Springer, 2008.

Ionized PVD and its pulsed variant HiPIMS are also of growing importance. An introduction to these topics can be found in these two references.

Jeffrey Hopwood, ed., *Ionized Physical Vapor Deposition*, Thin Films volume 27, Academic Press, 2000.

D. Lundin and K. Sarakinos, "An introduction to thin film processing using high-power impulse magnetron sputtering", *J. Mater. Res.* **27**, 780-792, 2012

APPENDIX A. MATHEMATICA DECODER

The table below explains the Mathematica syntax and functions needed to understand the code in this book. For a deeper explanation, you should consult the Mathematica documentation or one of the many books on the topic. I have found the books by Ruskeepaa (Ruskeepaa 2009) and Wellin (Wellin 2013) helpful.

Mathematica Code	Returns	Comment
a=2	2	Assign a value
a=2;		Semicolon suppresses output
A={5,6,7}	{5,6,7}	Assign a list of 3 elements to A
A[[1]]	5	Grab first element in list
A[[-1]]	7	Grab last element in list
A[[1;;2]]	{5,6}	Grab first 2 elements of list
Map[#+1&, A]	{6,7,8}	Map maps a function onto each element of a list. In this case each element (represented by #) has 1 added to it
B={{"a",5},{"b",6},{"c",7}};		Create a 2d list of mixed elements
Map[#[[2]]+1&,B]	{6,7,8}	
Flatten[B]	{"a",5,"b",6,"c",7}	Reduce the dimensionality of a list
func=Interpolation[{1,2},{2,3},{3,4}]	Interpolating function	Create a function that interpolates between points in a list
func[1.5]	2.5	Interpolating functions can be called like any other function
a+b /. {{a->1}, {b->5}}	6	The replaceAll operator (/.) is used to replace two symbols
B /. {a_, b_} -> a	{a,b,c}	Replacement rule in which any 2 element list is replaced with the first element

Code	Output	Description
p1=Plot[Sin[x],{x,0,pi}]		Plot the function Sin from 0 to pi
p2=Plot[Cos[x],{x,0,pi},ImageSize -> Large]		A plot of Cos[x], with one Plot's options invoked
Show[p1,p2]		Create a composite plot of p1 and p2
f[x_] := x+1		Define a function f which takes a parameter x and returns x+1. The "_" means x can be anything.
f[x_?NumericQ] := x+1		This function only accepts numeric arguments
f[x_] := Module[{a}, a=1; x+a]		For functions requiring local variables, Module is used.
NIntegrate[f[x],{x,0,2}]	4	Numerically integrate our function from x = 1 to 2
Sum[f[x],{x,1,10}]	65	Sum up f[x] for x = 1, 2...10
Range[0,1,0.1]	{0, 0.1, 0.2, ...}	Create a list of numbers
Clear[A]		Remove any definitions of A
NDSolve[{g'[x]==Sin[x],g[0]==1},g,{x,0,Pi}]	Interpolating function	Numerically solve an ordinary differential equation
Table[2*i,{i,1,5}]	{2,4,6,8,10}	The Table function can be used to generate lists

APPENDIX B. CROSS SECTIONS

e-Ar cross sections adapted from Hayashi
Energy in eV, cross sections in 10^-20 m^2

Energy	Elastic	Excitation	Ionization	Total
0	7.79	0	0	7.79
0.5	0.473	0	0	0.473
1	1.43	0	0	1.43
1.5	2.41	0	0	2.41
2	3.52	0	0	3.52
2.5	4.53	0	0	4.53
3	5.5	0	0	5.5
3.5	6.34	0	0	6.34
4	7.18	0	0	7.18
4.5	8.14	0	0	8.14
5	9.1	0	0	9.1
5.5	10.15	0	0	10.15
6	11.2	0	0	11.2
6.5	12.425	0	0	12.425
7	13.65	0	0	13.65
7.5	14.875	0	0	14.875
8	16.1	0	0	16.1
8.5	17.175	0	0	17.175
9	18.25	0	0	18.25
9.5	19.325	0	0	19.325
10	20.4	0	0	20.4
10.5	21.1	0	0	21.1
11	21.8	0	0	21.8
11.5	22.5	0	0	22.5
12	23.2	0.0164	0	23.2164
12.5	23.3	0.0377	0	23.3377
13	23.4	0.0544	0	23.4544
13.5	23.5	0.0765	0	23.5765
14	23.6	0.11	0	23.71
14.5	23.7	0.1386	0	23.8386
15	23.8	0.1622	0	23.9622
15.5	23.39	0.1827	0	23.5727
16	22.98	0.2033	0.0202	23.2035
16.5	22.57	0.2203	0.0771	22.8674
17	22.16	0.2374	0.134	22.5314
17.5	21.75	0.2488	0.214	22.2128
18	21.34	0.2603	0.294	21.8943
18.5	20.93	0.2684	0.377	21.5754
19	20.52	0.2766	0.46	21.2566
19.5	20.11	0.2843	0.5435	20.9378

20	19.7	0.292	0.627	20.619
25	15.5	0.3267	1.3	17.1267
30	12.5	0.3483	1.8	14.6483
35	10.815	0.3678	2.175	13.3578
40	9.13	0.385	2.39	11.905
45	8.255	0.3919	2.49	11.1369
50	7.38	0.3988	2.53	10.3088
60	6.34	0.3963	2.66	9.3963
70	5.765	0.3841	2.77	8.9191
80	5.19	0.3641	2.84	8.3941
90	4.905	0.3452	2.86	8.1102
100	4.62	0.3263	2.85	7.7963
150	3.7	0.2457	2.68	6.6257
200	3.13	0.2012	2.39	5.7212
250	2.77	0.1762	2.17	5.1162
300	2.5	0.1511	1.98	4.6311
350	2.32	0.1352	1.81	4.2652
400	2.14	0.1225	1.68	3.9425
450	2.015	0.1134	1.57	3.6984
500	1.89	0.1044	1.46	3.4544
600	1.71	0.0941	1.3	3.1041
700	1.58	0.0839	1.16	2.8239
800	1.45	0.0736	1.06	2.5836
900	1.355	0.0678	0.988	2.4109
1000	1.26	0.0621	0.916	2.2381
2000	1.12	0.0552	0.8142	1.9894
10000	0	0	0	0

APPENDIX C. SPEEDING UP MATHEMATICA

Two things make the Mathematica code in Chapter 6 slow.   First, Mathematica is an interpreted language.   So unlike C or Fortran, it must process each command at run time.   Second, we are calling NDSolve many times during each electron flight.   Each time it is called there is some setup involved.   The code below resolves both issues.   The Compile function is used to compile the Mathematica code to an intermediate language that the Wolfram Virtual Machine can run quickly.   Second, NDSolve is replaced with a fourth order Runge-Kutta solver, which is both fast and accurate (Press, et al. 2007).

```
rk4 = Compile[{{dt, _Real}, {t, _Real}, {x, _Real},
   {y, _Real}, {z, _Real}, {vx, _Real}, {vy, _Real}, {vz, _Real}},
   Module[{dt2, dt6, xVect, xVectOut, dxdt, xVectT, dxT, dxM, t1},

      xVect = {x, y, z, vx, vy, vz};
      dt2 = dt / 2.; dt6 = dt / 6.; t1 = t + dt2;

      dxdt = derivsY[t1, xVect];
      xVectT = xVect + dxdt * dt2;

      dxT = derivsY[t1, xVectT];
      xVectT = xVect + dxT * dt2;

      dxM = derivsY[t1, xVectT];
      xVectT = xVect + dxM * dt;
      dxM = dxM + dxT;

      dxT = derivsY[t1, xVectT];
      xVectOut = xVect + dt6 * (dxdt + dxT + 2. * dxM);
      xVectOut
   ],
   CompilationOptions -> {"InlineExternalDefinitions" -> True}
];
```

```
derivsY = Compile[{{t, _Real}, {xVect, _Real, 1}},

    Module[{pt, vx, vy, vz, ax, ay, az, bx = 0., by = 0., bz = 0., Ez,
      pt0 = {0., 0., 0.}, size, Br, X1, Y1, Z1, X2, Y2, Z2, numMags, i, z0},

      (*Set up position and vel vectors*)
      pt = xVect[[1 ;; 3]];
      vx = xVect[[4]]; vy = xVect[[5]]; vz = xVect[[6]];

      (*Loop over number of magnets in magpack*)
      numMags = Length[magpack2];
      For[i = 1, i ≤ numMags, i++,
        (*Get corner pt, size and strength of mag*)
        pt0 = magpack2[[i, 1]];
        size = magpack2[[i, 2]]; Br = magpack2[[i, 3, 1]];

        (*Calc all distances*)
        {X1, Y1, Z1} = pt - pt0;
        {X2, Y2, Z2} = pt - (pt0 + size);

        (*Calc the fields*)
```

$$bx = bx + \frac{-Br}{4. \times 3.1415926} \left(-Log\left[Y1 + \sqrt{X1^2 + Y1^2 + Z1^2}\right] + Log\left[Y1 + \sqrt{X2^2 + Y1^2 + Z1^2}\right]\right.$$
$$+ Log\left[Y2 + \sqrt{X1^2 + Y2^2 + Z1^2}\right] - Log\left[Y2 + \sqrt{X2^2 + Y2^2 + Z1^2}\right]$$
$$- Log\left[Y2 + \sqrt{X1^2 + Y2^2 + Z2^2}\right] + Log\left[Y2 + \sqrt{X2^2 + Y2^2 + Z2^2}\right]$$
$$\left.+ Log\left[Y1 + \sqrt{X1^2 + Y1^2 + Z2^2}\right] - Log\left[Y1 + \sqrt{X2^2 + Y1^2 + Z2^2}\right]\right);$$

$$by = by + \frac{-Br}{4. \times 3.1415926} \left(-Log\left[X1 + \sqrt{X1^2 + Y1^2 + Z1^2}\right] + Log\left[X2 + \sqrt{X2^2 + Y1^2 + Z1^2}\right]\right.$$
$$+ Log\left[X1 + \sqrt{X1^2 + Y2^2 + Z1^2}\right] - Log\left[X2 + \sqrt{X2^2 + Y2^2 + Z1^2}\right]$$
$$- Log\left[X1 + \sqrt{X1^2 + Y2^2 + Z2^2}\right] + Log\left[X2 + \sqrt{X2^2 + Y2^2 + Z2^2}\right]$$
$$\left.+ Log\left[X1 + \sqrt{X1^2 + Y1^2 + Z2^2}\right] - Log\left[X2 + \sqrt{X2^2 + Y1^2 + Z2^2}\right]\right);$$

```
        bz =
```

$$bz + \frac{Br}{4. \times 3.1415926} \left(-ArcTan\left[\frac{X1\,Y1}{Z1\sqrt{X1^2 + Y1^2 + Z1^2}}\right] + ArcTan\left[\frac{X2\,Y1}{Z1\sqrt{X2^2 + Y1^2 + Z1^2}}\right]\right.$$
$$+ ArcTan\left[\frac{X1\,Y2}{Z1\sqrt{X1^2 + Y2^2 + Z1^2}}\right] - ArcTan\left[\frac{X2\,Y2}{Z1\sqrt{X2^2 + Y2^2 + Z1^2}}\right]$$
$$- ArcTan\left[\frac{X1\,Y2}{Z2\sqrt{X1^2 + Y2^2 + Z2^2}}\right] + ArcTan\left[\frac{X2\,Y2}{Z2\sqrt{X2^2 + Y2^2 + Z2^2}}\right]$$
$$\left.+ ArcTan\left[\frac{X1\,Y1}{Z2\sqrt{X1^2 + Y1^2 + Z2^2}}\right] - ArcTan\left[\frac{X2\,Y1}{Z2\sqrt{X2^2 + Y1^2 + Z2^2}}\right]\right);$$

```
      ];
      z0 = pt[[3]];
```

$$Ez = If\left[z0 < s, \frac{2\,Vd}{s} \frac{s - z0}{s}, 0\right];$$

```
      (*Velocity derivatives*)
      ax = -q / m * (vy * bz - vz * by);
      ay = -q / m * (-vx * bz + vz * bx);
      az = -q / m * (Ez + vx * by - vy * bx);

      (*Return array of derivatives*)
      {vx, vy, vz, ax, ay, az}

    ]
  ];
```

```
moveItFast[x0_, v0_, tstep_] := Module[{x1, v1, vect, t = 0, dt = 3 × 10^-12},
   (*Print["MoveIt:v0 = ",v0];*)
   x1 = x0; v1 = v0;
   While[t < tstep,
     (*Print["x, v = ",x1,v1];*)
     (*vect=rk4[1 10^-12,t,x1,v1];*)
     vect = Apply[rk4, Join[{dt, t}, x1, v1]];
     (*Print[vect];*)
     x1 = vect[[1 ;; 3]]; v1 = vect[[4 ;; 6]];

     (*vList=Append[vList,vect];*)
     t = t + dt;
   ];

   (*Return the position and velocity at t=tstep*)
   Return[{x1, v1}];

];
```

```
runE[p0_, v0_, tmax_, tstep_] :=

 Module[{time, p1, p2, v1, v2, delx, KE, prob, r, type = 0, vMag,
   vMag2, plist = {p0}, vlist = {v0}, eList = {}, collList = {}},
  time = 0.0;
  p1 = p0; v1 = v0;
  (*Print["time, tmax =",time,tmax];
  Print["totEn= ",totEnergy[p1,v1]];*)
  While[time < tmax && totEnergy[p1, v1] > 16. && p1[[1]] > -0.10,
   (*Print["totEn= ",totEnergy[p1,v1]];*)
   {p2, v2} = moveItFast[p1, v1, tstep];

   delx = EuclideanDistance[p1, p2];
   vMag = Norm[v2];
```

$$KE = 0.5\, m\, vMag^2\ \frac{1\ (*eV*)}{1.602 \times 10^{-19}\ (*J*)};$$

```
   prob = collProb[nGas, KE, delx];
   (*Print["delx, KE, prob = ", delx," ",KE, " ", prob];*)
   r = RandomReal[];
   If[ r < prob,
    type = decideEvent[KE];
    v2 = newVel[v1, KE, type];
    {collList, eList} = addPtToList[p2, type, collList, eList];
    ];
   (*Update position and velocity lists for time step*)
   plist = Append[plist, p2];
   vlist = Append[vlist, v2];

   (*Update time and position*)
   time = time + tstep;
   p1 = p2; v1 = v2;
   ]; (*while loop*)
  Return[{plist, vlist, collList, eList}];

 ]
```

REFERENCES

BIBLIOGRAPHY

Ansys, Inc. *Ansys Maxwell.* Inc. Ansys. 2014. http://www.ansys.com/ (accessed March 1, 2014).

Bohm, David. *The Characteristics of Electrical Discharges in Magnetic Fields.* Edited by A. Gurthrie and K. Wakerling. New York, NY: McGraw-Hill, 1949.

Bradley, J. W., S. Thompson, and Y. A. Gonzalvo. "Measurement of the plasma potential in a magnetron discharge and the prediction of the electron drift speeds." *Plasma Sources Science and Technology* 10 (2001): 490-501.

Bultinck, E., S. Mahieu, D. Depla, and A. Bogaerts. "The origin of Bohm diffusion, investigated by a comparison of different modeling methods." *Journal of Physics D* 43 (2010): 292001.

Buyle, G., K. Depla, K. Eufinger, J. Haemers, R. De Gryse, and W. De Bosscher. "Characterization of the Electron Movement in Varying Magnetic Fields and the Resulting Anomalous Erosion ." *47th Annual Technical Conference Proceedings.* Society of Vacuum Coaters, 2004. 265-270.

Buyle, Guy. "Simplified Model for the DC Planar Magnetron Discharge." PhD Dissertation, Ghent University, 2005.

Chen, Francis F. *Introduction to Plasma Physics and Controlled Fusion. Vol 1: Plasma Physics.* 2nd Edition. New York, NY: Plenum Press, 1984.

Choi, Young Wook, Mark Bowden, and Katsunori Muraoka . "A Study of Sheath Electric Fields in Planar Magnetron Discharges using Laser Induced Fluorescence Spectroscopy." *Japanese Journal of Applied Physics* 35 (1996): 5858-5861 .

References

Cramer, N. F. "Analysis of a one-dimensional, steady-state magnetron discharge ." *Journal of Physics D* 30 (1997): 2573-2584.

Dexter Magnetic Technologies. *Dexter Magnetic Technologies.* 2014. http://www.dextermag.com/ (accessed March 1, 2014).

Engel-Herbert, R., and T. Hesjedal . "Calculation of the magnetic stray field of a uniaxial magnetic domain ." *Journal of Applied Physics* 97 (2005): 074504.

Fan, Q. H., L. Q. Zhou, and J. J. Gracio. "A cross-corner effect in a rectangular sputtering magnetron." *Journal of Physics D* (IOP) 36 (2003): 244–251.

Grove, W. R. "On the Electro-Chemical Polarity of Gases ." *Philosophical Transactions of the Royal Society* (Royal Society of London) 142 (1852): 87-101.

Gu, Lan, and Michael Lieberman. *Journal of Vacuum Science and Technology A* 6, no. 5 (Sep/Oct 1988): 2960-2964.

Hayashi. *Argon-Electron Cross Sections with 25 Excited States.* 2005. http://jila.colorado.edu/~avp/collision_data/electronneutral/hayashi.txt (accessed March 1, 2014).

Integrated Engineering Software. *Lorentz.* Integrated Engineering Software. 2014. http://www.integratedsoft.com/ (accessed March 1, 2014).

Jackson, John David. *Classical Electrodynamics.* 2nd Edition. New York, NY: John Wiley & Sons, 1975.

Liberman, Michael A., and Alan J. Lichtenberg. *Principles of Plasma Discharges and Materials Processing.* 2nd Edition. New York, NY: Wiley-Interscience, 2005.

Lister, Graeme. "Influence of electron diffusion on the cathode sheath of a magnetron discharge ." *Journal of Vacuum Technology A* 14, no. 5 (1996): 2736-2743.

Musschoot, J, D Depla, G Buyle, and R DeGryse. "Investigation of the sustaining mechanisms of dc magnetron discharges and consequences for I–V characteristics ." *Journal of Physics D* 41 (2008): 015209.

Okhrimovskyy , A., A. Bogaerts , and R. Gijbels . "Electron anisotropic scattering in gases: A formula for Monte Carlo simulations." *Physical Review E* 65 (2002): 037402 .

Popkov, Alexey. *Mathematica Stack Exchange*. September 9, 2013. http://mathematica.stackexchange.com/questions/31858/how-to-extract-data-from-contour-plot-as-a-text-file (accessed August 6, 2014).

Press, William H., Saul A. Teukolsky, William T. Vetterling, and Brian P. Flannery. *Numerical Recipes*. 3rd Edition. New York, NY: Cambridge University Press, 2007.

Ruskeepaa, Heikki. *Mathematica Navigator*. 3rd Edition. Amsterdam: Academic Press, 2009.

Sheridan, T. E., M. J. Goeckner, and J. Goree. "Model of energetic electron transport in magnetron discharges." *Journal of Vacuum Technology A* 8, no. 1 (1990): 30-37.

Sheridan, Terrence E., and John A. Goree. "Analytic Expression For the Electric Potential in the Plasma Sheath." *IEEE Transactions on Plasma Science* 17, no. 6 (December 1989): 884-888.

Siegel, Robert, and John R Howell. *Thermal Radiation Heat Transfer*. 3rd Edition. Washington DC: Hemisphere Publishing Corporation, 1992.

Wellin, Paul. *Programming with Mathematica* . New York: Cambridge University Press, 2013.

Wikipedia. *Mars Climate Orbiter*. 2104. http://en.wikipedia.org/wiki/Mars_Climate_Orbiter (accessed March 1, 2014).

Yanguas-Gil, Angel, Jose Cotrino, and Luis L Alves. "An update of argon inelastic cross sections for plasma discharges." *Journal of Physics D* 38 (2005): 1588-1598.

INDEX

anomalous diffusion, 112

argon, x, 1, 3, 5, 46, 72, 81, 82, 83, 85, 87, 89, 90, 91, 102, 110, 113, 114, 117, 128

B field, vii, 4, 5, 6, 9, 10, 11, 13, 15, 17, 18, 19, 20, 22, 23, 24, 25, 27, 28, 31, 50, 51, 53, 54, 55, 58, 59, 65, 67, 68, 69, 71, 72, 73, 74, 75, 77, 79, 80, 81, 96, 97, 98, 106, 121, 129

Bohm diffusion, 96, 115

Bohm velocity, 46

chemical vapor deposition, 2

collisions, 81, 82, 83, 87, 89, 90, 91, 94, 95, 96, 99, 100, 101, 102, 104, 128

elastic, 81, 82, 85, 86, 87, 89, 90, 99, 101, 102

ionization, 7, 37, 72, 74, 77, 81, 82, 83, 84, 86, 87, 89, 90, 99, 101, 102, 109, 110, 112, 113, 115

cosine distribution, 118

cross corner effect, 74, 78

cross section, 82, 83, 84, 85, 86, 87, 117, 133

curvature drift, 66, 67, 68, 69, 70

CVD, 2

cyclotron, 6, 96

deposition rate, x, 2, 123, 124, 125, 126, 128

drift velocity, 4, 24, 64, 65, 66, 67, 68, 70, 72, 73, 74, 81

electric field, 1, 3, 6, 33, 34, 35, 37, 43, 49, 51, 56, 57, 59, 96, 129

electron, x, 1, 3, 4, 5, 6, 7, 9, 15, 34, 43, 44, 50, 53, 54, 57, 58, 59, 60, 61, 62, 63, 65, 66, 67, 68, 69, 70, 71, 72, 73, 74, 75, 76, 77, 78, 79, 80, 81, 82, 83, 84, 85, 87, 89, 90, 91, 92, 93, 94, 95, 96, 97, 98, 99, 100, 101, 102, 103, 104, 105, 106, 107, 109, 110, 113, 114, 115, 129, 135

energy, 1, 3, 33, 37, 43, 59, 60, 61, 62, 72, 81, 85, 87, 89, 90, 91, 92, 95, 96, 99, 101, 102, 105, 106, 107, 108, 110, 113, 128

ionization, 102

kinetic, 5, 37, 59, 60, 61, 78, 91, 101

loss, 90

erosion, x, 7, 37, 39, 72, 74, 77, 78, 80, 96, 99, 108, 111, 113, 114, 118, 120, 128, 129

ExB drift, 22, 59, 66, 67, 68, 69, 70, 77, 78, 106

Fortran, 109, 114, 135

grad B drift, 67

Graphics Complex, 121

Grove, W.R., ix

gyroradius, 4, 50

HiPIMS, 130

interpolation, 85

ion, 1, 34, 35, 36, 37, 38, 43, 46, 47, 48, 51, 99, 102, 129

Larmor radius, 4, 50, 67, 106

Lorentz force, 3, 50, 53

magnet array, 18, 21, 22, 26, 27, 28, 30, 31, 54, 75, 76, 77, 78, 103, 104, 108, 109, 113, 117

magnetic field, x, 3, 9, 10, 13, 19, 31, 50, 53, 65, 70, 75, 84, 96, 117, 129

magnetization, 10, 11, 12, 54

magnetron, 3, 4, 5, 6, 9, 15, 25, 27, 28, 30, 37, 50, 59, 64, 75, 80, 84, 114, 129, 130

Map, 19, 85, 109, 111, 112, 122, 131

Maple, ix

Mathematica, viii, ix, 13, 15, 16, 18, 25, 40, 44, 45, 53, 54, 55, 56, 58, 59, 85, 94, 95, 99, 109, 120, 121, 131, 135

Matlab, ix

Monte Carlo, 37, 50, 112, 128

NDSolve, 44, 54, 56, 57, 61, 64, 71, 76, 100, 109, 132, 135

NumericQ, 132

orbit, 6, 65, 74, 129

parametric plot, 58

physical vapor deposition, ix, 129, 130

Poisson's equation, 11, 34, 38, 44, 45, 50

pole plate, 9, 29, 30, 31

potential, 10, 11, 29, 34, 35, 37, 39, 40, 42, 44, 45, 48, 50, 59, 60, 61, 72

precision, 45, 58, 120

probability, 3, 82, 83, 87, 89, 90, 92, 96, 98, 101

PVD, vii, ix, x, 2, 3, 10, 34, 81, 83, 84, 87, 129
    ionized, 130

racetrack, 5, 6, 22, 24, 25, 26, 27, 28, 39, 46, 50, 51, 64, 66, 70, 76, 77, 78, 79, 110, 112, 113, 118, 120, 121, 122, 128

radiation analogy, viii, 118

random number, 92, 101, 102

remanence, 12

Runge-Kutta, 109, 135

scatter, 82, 92, 96, 97, 98, 112, 117

semiconductor, 129

sheath, 4, 34, 35, 36, 37, 38, 39, 40, 42, 43, 45, 46, 48, 49, 50, 51, 56, 57, 59, 67, 72, 102, 104, 113, 129
    ion matrix, 35, 36, 37, 46, 51
    two fluid, 44, 45, 46, 48, 49

speedup, 109

sputter, x, 33, 53, 72, 102, 109, 114, 117, 120, 128

sputtering, ix, x, 2, 3, 5, 6, 9, 10, 33, 34, 51, 64, 72, 77, 81, 84, 90, 99, 102, 114, 120, 128, 129, 130
    reactive, 129

target, x, 1, 2, 3, 4, 5, 7, 9, 15, 17, 22, 23, 24, 25, 27, 28, 31, 33, 36, 37, 38, 39, 44, 45, 50, 53, 54, 55, 56, 57, 58, 59, 61, 62, 63, 64, 71, 72, 74, 76, 77, 78, 79, 81, 84, 99, 102, 104, 106, 108, 109, 111, 112, 113, 114, 115, 117, 118, 120, 121, 122, 123, 125, 126, 128, 129
    utilization, 24, 72, 112

turnaround, 23, 70, 74, 76, 77, 78, 79, 80, 120

uniformity, 7, 117, 125, 128

units, 5, 82, 83, 85, 123, 124

view factor, 117, 118, 119, 123, 128

Wolfram Virtual Machine, 135

ABOUT THE AUTHOR

Jack McInerney is a senior staff scientist in the Computational Modeling group of Lam Research. His computational explorations have ranged broadly from barbecuing meat to plasma physics. He has authored over 30 technical papers and holds more than a dozen patents.

Made in the USA
San Bernardino, CA
08 February 2016